CROSS-BEDDING, BEDFORMS, AND PALEOCURRENTS

David M. Rubin

U.S. Geological Survey, Menlo Park, California

SOCIETY OF ECONOMIC PALEONTOLOGISTS AND MINERALOGISTS

Concepts in Sedimentology and Paleontology, Volume 1

Barbara H. Lidz, Editor of Special Publications

You don't need a weatherman
to know which way the wind blows.

—Bob Dylan
Subterranean Homesick Blues

CONTENTS

FOREWORD

The geometry of stratification produced by the changing and shifting of bedforms during deposition is almost always complexly three-dimensional, but we too often lose sight of this when we look at stratification shown on rock surfaces. The view on such surfaces is likely to be highly misleading, and even seeing two or three different sections normal to bedding usually leaves plenty of room for ambiguity. Even if it were practical to do serial sectioning of whole outcrops, there would still be the problem that information about bed geometry is lost whenever there is local erosion as well as deposition while the bedforms shift, as is common in cross-stratified deposits. How, then, can we deduce bed geometry from the preserved record of stratification?

Rubin has attacked the problem from the opposite direction with a simple but fruitful idea: by assuming reasonable bed geometry and letting it change with time during deposition, he creates the stratification, in the form of gorgeous three-dimensional block diagrams plotted by computer. This magnificent catalog, along with its intelligent commentary, should serve to guide our thinking about the relationship between bed geometry and stratification for years to come. One could argue that many of the examples are too regular—but reference cases are needed to make some sense of the infinity of possible cases—or one could argue that some of the assumed bed geometries don't develop in nature. That is bound to be true in an exercise like this, but a great many of the pictures in this book should strike a chord of recognition in the minds of geologists who have looked at cross-stratification, and therein will lie the value of this imaginative book.

John Southard
Massachusetts Institute of Technology
Cambridge, Massachusetts

PREFACE

The computer modeling that forms the basis for this publication was undertaken to relate the geometry of cross-bedding to the morphology and behavior of bedforms. Using computers for this purpose is necessary only because sedimentologists cannot adequately visualize the geometry arising when complex, changing surfaces move through space. People differ in their ability to visualize such geometric processes, and a few of the images in this collection will undoubtedly be too simple to be of general interest; other images may be too complicated to be comprehended thoroughly by anyone. Although the details in many of the complicated images will be of interest only to sedimentologists who have had experience interpreting cross-bedding or studying the behavior of bedforms, I hope that the simpler illustrations can convey an understanding of the origin of cross-bedding geometry even to those not experienced in the field.

This publication is only an introduction to the technique of computer modeling of cross-bedding. To take full advantage of this powerful technique requires personal experience at the trial-and-error simulation of structures observed in the field. This trial-and-error experimentation leads to interpretations that precisely recreate bedform behavior and morphology. Moreover, such experience conveys an intuitive understanding of how the various bedform-description parameters influence the resulting bedding. Perhaps within the next decade small portable computers will be capable of simulating cross-bedding at the outcrop in a minute or two of operation.

Ralph Hunter (U.S. Geological Survey, Menlo Park) interested me in the origin of cross-bedding. He spent many weeks in the field with me discussing problems of cross-bedding interpretation, and many of the ideas in this publication evolved out of our joint work. Rex Sanders (U.S. Geological Survey, Menlo Park) provided advice about FORTRAN programming and about plotters and computer systems operation. Jack Schmidt (U.S. Geological Survey, Tucson) provided the opportunity to work with him on the Colorado River in the Grand Canyon. Johnnie Moore (University of Montana) and Tom Clifton (University of California at Santa Cruz) helped in the trenching and examination of Colorado River deposits. Bob Dalrymple (Queen's University, Kingston), Ralph Hunter, Dave McCulloch (U.S. Geological Survey, Menlo Park) and John Southard (Massachusetts Institute of Technology, Cambridge) reviewed the manuscript for this publication.

INTRODUCTION

Cross-bedding analysis is an indispensable technique for interpreting sedimentary deposits. Cross-bedding is an important indicator of depositional environments, paleoflow velocities, and paleocurrent directions. Although most studies of cross-bedding have been directed toward determining ancient flow conditions or toward basic research to enable such applied studies, a second reason to study cross-bedding is equally important. Cross-bedding provides information about how modern bedforms behave—how they migrate, how they change in morphology, and how they interact with other bedforms. This kind of information is particularly valuable for bedforms whose behavior cannot be observed, either because they migrate or change shape too slowly, or because they are active in environments where repeated observations cannot be made. Such studies of cross-bedding should be of particular interest to fluid dynamicists who are developing models of interactions between bedforms and flow, because bedforms and the flows that produce them cannot be modeled accurately until the behavior of bedforms is known. For example, the tendency of "longitudinal" dunes to migrate laterally was recognized from studies of eolian cross-bedding before modern dunes were known to exhibit such behavior (Rubin and Hunter, 1985). Subsequently, sedimentologists have begun to wonder why bedforms behave in such a manner.

Regardless of whether the goal is to interpret sedimentary deposits or to learn how bedforms behave, interpretation of cross-bedding is a two-step process. The first step—reconstructing the morphology and behavior of bedforms from cross-bedding—is primarily a problem of solid geometry. This geometrical problem forms the basis of this publication. The second step—determining flow conditions from bedform morphology and behavior—is mainly a problem of fluid dynamics. Previous studies of the origin of cross-bedding have not always adequately distinguished these two steps but have nevertheless greatly advanced our understanding of the origin of cross-bedding. The conceptual breakthrough regarding the origin of cross-bedding was Sorby's (1859) realization that cross-laminated beds originate by the climbing of ripples. Subsequent studies have relied on a variety of approaches, including field observations of bedforms and their internal structures, flume experiments, theoretical analyses, and inferences based on observations of stratification in rocks.

In general, previous studies of cross-bedding geometry have had three limitations. First, they have been restricted in scope. Studies have usually been directed at determining the origin of one particular structure or group of related structures, determining what structures are produced by a specific bedform or produced in a specific environment, or determining what structures arise from a specific process such as reversing flow or migration of superimposed bedforms. As a result, some common structures have not been studied. Second, virtually all previous studies of cross-bedding have treated bedforms as quasi-two-dimensional features. Even where bedforms have been considered to be three-dimensional, migration of the three-dimensional features such as scour pits or plan-form sinuousities has been treated as occurring only in a direction normal to the bedform crestline (in the case of transverse bedforms) or parallel to the crestline (in the case of longitudinal bedforms). The existence of oblique bedforms with both transverse and longitudinal components of sediment transport has usually been ignored. Third, the results of previous studies have usually been presented in sketches that are limited by the investigators' ability to visualize and draw in three dimensions. Consequently, illustrations have not always been accurate and sometimes have shown bedforms with internal structures that the depicted bedforms could not have produced.

The aim of this publication is to present a collection of computer-generated images of cross-bedding that is broad in scope, that includes models of bedforms that behave in a more realistic manner than most previous conceptions, and in which the individual images accurately depict the depositional situations that are modeled. Computer-graphics modeling is a new tool that is ideal for this purely geometric problem of relating bedforms to cross-bedding. Computers can be used to create bedforms, to cause them to migrate, and to display the internal structures resulting from bedform migration. The resulting images are powerful instructive tools because the major aspects of deposition are determined by the experimenter, and, consequently, the origin of the structures is not subject to question. The images are also instructive because the origin of the structures can be related to specific bedforms, the movement and behavior of those bedforms can be depicted through time, a wide variety of cross-bedded structures can be produced, outcrop orientation can be selected, and illustrations accurately depict the mathematically generated bedding. The computer images are compared with a smaller number of field photographs to show that the computer model is simulating processes that occur in the real world.

COMPUTER MODEL

Operation

Each of the computer images in this publication is the result of a geometry experiment conducted by computer. For each experiment, the computer is given a list of parameters that define the morphology and behavior of a particular bedform or assemblage of bedforms. The parameters specified for each experiment include the spacing, steepness, asymmetry, migration direction, migration speed, planform shape, and along-crest migration speed of planform sinuosities of each set of bedforms (Appendix A). Because bedform morphology and behavior can vary through time, the computer model was designed to be capable of varying most of these parameters cyclically through time; the magnitude, period, and phase of variations in any changing parameter must be specified. Specifying these parameters for as many as three sets of simultaneously or alternately active bedforms, specifying how the various bedforms are to be superimposed, specifying the rate of deposition, and specifying changes through time in the rate of deposition requires 75 geometric parameters.

With these input specifications, the computer program uses sine curves to create mathematically surfaces that approximate the shape of bedforms. Displacement of the sine curves simulates bedform migration, changing amplitude simulates changing bedform height, and combining separate sets of sine curves simulates superpositioning of bedforms. The program is capable of combining the curves according to several different rules: plain addition, adding a percentage of the superimposed curve that is either proportional or inversely proportional to the local elevation of the main curve, or selecting the curve that locally has the greatest elevation (Appendix A). All rules except the latter produce bedforms that generate foresets with tangential basal contacts; the latter rule produces foresets with angular basal contacts.

Three computer programs were used to produce the images in this publication; by varying only the input parameters, each program models different depositional situations. The three programs all use the same equations to define the bed surface, but the results are displayed differently. The first program calculates the topography of the bed surface and displays that surface in three-dimensional perspective. The program then migrates the bedforms backward through time and space, distinguishes preserved surfaces from nonpreserved surfaces, and plots the traces of the older preserved surfaces on vertical outcrop planes. The resulting images include both bed morphology and internal structures and are therefore useful for relating bedforms to bedding (Fig. 1).

The second program produces perspective block diagrams with horizontal sections instead of bed morphology at the top of the block (Fig. 1). The horizontal sections are generated using a contour-mapping program that contours a single elevation at different instants in time, rather than the usual contours of multiple elevations at a single time. In addition to connecting the bedding traces visible in the two vertical sections, the horizontal sections are useful for illustrating such features as cross-bed strike, scour-pit paths, and plan-form shape of the bedforms.

The third program plots vectors that represent the migration of bedforms and scour pits, it plots the direction of sediment transport represented by bedform migration azimuth, and it plots inclination of cross-bed and bounding-surface planes (Fig. 1). The plotted migration vectors are the mean vectors; the cyclic variations in migration vectors that are present in some of the simulated depositional situations are incorporated in the mean value but are not plotted individually. The same program randomly selects points on the bed surface and then calculates the azimuth and inclination of cross-beds and bounding surfaces that occur in vertical profiles beneath the randomly selected points. The dips are plotted with distance from the center of the plot proportional to the inclination of the bed and with azimuth indicated by the direction of the point from the center of the plot. In these plots, actual values were not assigned to inclinations to avoid having inclinations greater than the angle of repose; such steep angles might arise because the computer model does not incorporate physical processes such as avalanching or grainfall. The computer vertically profiles the structure resulting from each depositional situation at many locations, because the bedding commonly varies dramatically from one location to another in a single structure. Many vertical profiles are required to insure that the plot reflects the structure as a whole.

The plots resulting from this polar-plot program are displayed in a format that is commonly used for plotting randomly collected field measurements, but the nonrandom vertical profiling through the bedding structures is obvious in many of the plots. For example, in structures where the strike of the foresets is constant with depth in any set of cross-beds, all foresets sampled in a single vertical section plot along a line radiating from the center of the plot (row 3 in Fig. 1). When the structure is sampled at different locations, the foresets sampled in each section plot along a different radiating line, and collectively the points plot in a fan-shaped pattern.

CLASSIFICATION OF BEDFORMS AND CROSS-BEDDING

Advantages and Disadvantages

This computer model is purely geometric and does not incorporate theoretical fluid dynamics or empirical relations between flow conditions and bedforms. Although such considerations are crucial to the problem of relating bedforms to flow, fluid dynamics is largely irrelevant to the purely geometric problem of relating the morphology and behavior of bedforms to cross-bedding; however, even the second interpretive step of cross-bedding interpretation—relating bedform morphology and behavior to flow properties—is advanced somewhat by this geometric model. Specifically, displacement of bedforms and their superimposed topographic features can be used to determine the ratio of across-crest to along-crest sediment transport, thereby allowing determination of the sediment transport direction relative to the bedform crestline, which in turn defines the simulated bedforms as transverse, oblique, or longitudinal bedforms.

The main deficiency of this kind of geometric model is that the simulated bedform morphology and behavior are not constrained by physical processes. As a result, the model is capable of creating depositional situations that are physically impossible. Although such may prove to be the case with details in some of the images in this publication, the consequences of such an error are not great: merely that the computer-generated structure will not serve as an example for the origin of real structures, because no such real structures exist.

In addition to this general deficiency of any purely geometric model of cross-bedding, this particular computer model has several specific limitations. The simulated bedforms are more uniform (in directions both normal and parallel to their crestlines) than most real bedforms, and the bedforms behave in a more regular manner (individual bedforms do not split or merge with others). As a result, the simulated cross-bedding is more regular than most real cross-bedding. Rather than being a hindrance, however, the regularity of the computer-generated cross-bedding is beneficial, at least for the instructive purposes of this publication, because the simplified depositional situations are easier to visualize and understand. Some of the more random bedform properties are treated separately in a few images, so that the effects can be observed without obscuring the more regular and more comprehensible bedding features present in each image.

Approach

This publication uses a new classification scheme that relates the geometry of cross-bedding directly to the morphology and behavior of the bedforms that deposited the beds (Fig. 1). This scheme was developed because existing classifications of bedforms are generally not applicable to cross-bedding, and because existing classifications of cross-bedding do not adequately relate bedding geometry to bedform behavior. The approach toward both the modeling and classification in this publication emphasizes the shape and behavior of bedforms rather than size, flow, or fluid medium, and, consequently, bedforms are not subdivided into such categories as ripples, dunes, or sand waves, and the term "bedform" is used in a broad sense that includes all cyclic topographic features.

Most classifications of modern bedforms cannot be applied to cross-bedding because the morphologic and behavioral properties that can be determined from observation of bedforms are significantly different from the properties that can be determined from cross-bedding. Specifically, instantaneous observation of bedforms gives a detailed view of morphologic properties such as height, spacing, asymmetry, crestline sinuosity, and trough profile, but gives no indication of changes through time in bedform morphology or of transport-related characteristics such as the relative migration speeds of the main bedforms and superimposed bedforms, spurs, or scour pits. In contrast, cross-bedding commonly contains less information about the morphology of bedforms that existed at any one time but contains more information about morphologic history and transport-related behavior of bedforms. Existing bedform classification schemes generally cannot be applied to ancient bedforms because the ancient morphology is usually too imprecisely known, and, even in those cases where the classification schemes are imprecise enough to be applied or where the deposits are exceptionally revealing, the bedform morphologic history and behavior cannot be included in the classification.

Similarly, existing cross-bedding classifications have overlooked observable features that relate to bedform morphology and behavior and instead emphasize features that depend as much on outcrop orientation as on bedding geometry. For example, such classifications do not consider divergence in direction of dip between cross-beds and bounding surfaces, a feature which is included here because it is a

TWO-DIMENSIONAL Two-dimensional bedforms are straight and parallel in plan form; the flanks of the bedforms have the same strike at all locations. Two-dimensional bedforms produce two-dimensional cross-bedding: cross-bedding in which all foresets and bounding surfaces have the same strike. In plots showing the direction and inclination of dips of cross-beds and bounding surfaces, dips of all planes plot along a single straight line through the center of the plot.	**INVARIABLE** Invariable bedforms are those that do not change in morphology or path of climb. Cross-bedding deposited by invariable two-dimensional bedforms has bounding surfaces that are parallel planes; their poles plot as a single point.	**TRANSVERSE, OBLIQUE, AND LONGITUDINAL** Transverse, oblique, and longitudinal cross-bedding are not distinguishable unless bedforms are at least slightly three-dimensional (see below).
	VARIABLE Variable bedforms are those that change in morphology or path of climb. Variability causes dispersion in the inclination of bounding surfaces. Cross-bedding deposited by variable two-dimensional bedforms has bounding surfaces with a constant strike but with varying inclination; their poles plot as a straight line that parallels the line of cross-bed dips.	**TRANSVERSE, OBLIQUE, AND LONGITUDINAL** Transverse, oblique, and longitudinal cross-bedding are not distinguishable unless bedforms are at least slightly three-dimensional (see below).
THREE-DIMENSIONAL Three-dimensional bedforms are curved in plan form or have plan-form complexities such as scour pits or superimposed bedforms with a different trend from the main bedform; the strike of the flanks varies with location. Three-dimensional bedforms produce three-dimensional cross-bedding: cross-bedding in which foreset and bounding-surface strikes vary with location; dips of foresets do not plot along a single straight line through the center of polar plots.	**INVARIABLE** Cross-bedding deposited by invariable three-dimensional bedforms has bounding surfaces that are trough-shaped; bounding-surface dips in a single trough (or in identical troughs) plot as a nearly straight line.	**PERFECTLY TRANSVERSE** Plots of cross-bed and bounding-surface dips have bilateral symmetry; the axis of symmetry is the same for both plots; dip directions are distributed unimodally.
		OBLIQUE, IMPERFECTLY TRANSVERSE, OR IMPERFECTLY LONGITUDINAL Plots of cross-bed and bounding-surface dips do not have bilateral symmetry; cross-bed dips are asymmetrically distributed relative to bounding-surface dips.
		PERFECTLY LONGITUDINAL Plots of cross-bed and bounding-surface dips have bilateral symmetry; dip directions may be distributed bimodally (as shown) or may be unimodal as a result of migration of the nose of the main bedform. Perfect longitudinality is evidenced by vertical accretion of bedforms; cross-beds dip in opposing directions on opposite flanks.
	VARIABLE Bounding surfaces have complex shapes produced by such processes as zig-zagging of scour pits; dips of bounding surfaces plot as scatter diagrams.	**PERFECTLY TRANSVERSE** Same as perfectly transverse, invariable, three-dimensional cross-bedding.
		OBLIQUE, IMPERFECTLY TRANSVERSE, OR IMPERFECTLY LONGITUDINAL Same as oblique or imperfectly aligned, invariable, three-dimensional cross-bedding.
		PERFECTLY LONGITUDINAL Same as perfectly longitudinal, invariable, three-dimensional cross-bedding.

FIG. 1.— Scheme used to classify bedforms and organize the depositional situations and structures in this publication. From left to right the first three vertical columns define the classification parameters: three-dimensionality, variability, and orientation relative to transport. Column 4 shows block diagrams of

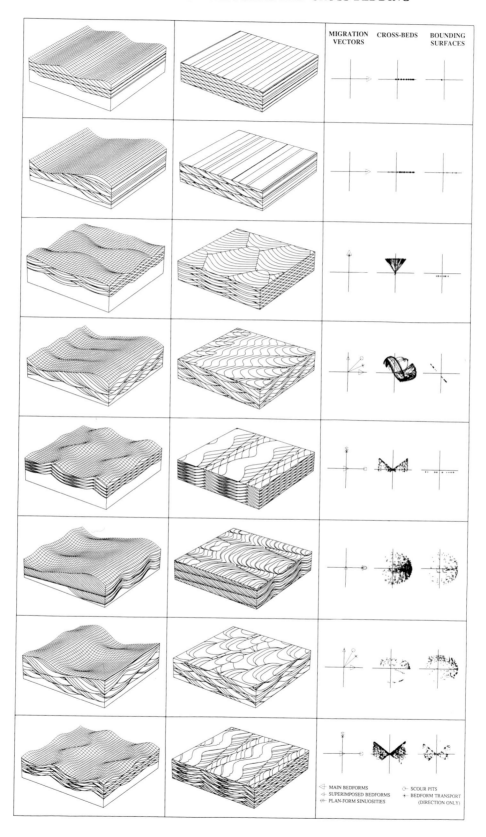

bedform morphology and vertical sections, column 5 shows block diagrams with horizontal and vertical sections, and column 6 shows polar plots of cross-bed and bounding-surface dip directions. The examples in the horizontal rows are simplifications of Figures 5, 17, 34A, 46H, 55, 59, 71, and 77.

key for distinguishing the deposits of transverse, oblique, and longitudinal bedforms.

The approach in this publication is to group cross-bedding and ancient bedforms into four main classes: (1) invariable two-dimensional bedforms and cross-bedding (illustrated in Figs. 2-11 with computer images and real examples that show salient depositional features), (2) variable two-dimensional bedforms and cross-bedding (Figs. 12-30), (3) invariable three-dimensional bedforms and cross-bedding (Figs. 31-56), and (4) variable three-dimensional bedforms and cross-bedding (Figs. 57-79). Bedforms can be further subdivided into transverse, oblique, and longitudinal categories depending on bedform orientation relative to the resultant sediment transport direction. The resulting classification groups bedforms by morphology (two-dimensional or three-dimensional), variability of bedform morphology and migration through time and space (invariable or variable), and bedform trend relative to the transport direction (transverse, oblique, or longitudinal). The same classification scheme can be applied to cross-bedding, because each of these classes of bedforms produces a distinctive kind of structure. Consequently, classifying the bedding simultaneously describes bedform morphology and behavior. In effect, classifying cross-bedding using this scheme is a first step in the interpretive process, rather than an end in itself. The following discussion explains the meaning of these classes, explains how to recognize examples of each class, and discusses the general properties of the bedforms required to produce structures of each class.

Terminology

Two-dimensional bedforms are defined as bedforms with straight crestlines, constant crest and trough elevations relative to the generalized depositional surface, and identical across-crest profiles at all locations along the crestline. Two-dimensional bedforms deposit two-dimensional cross-bedding—cross-bedding in which all foresets and bounding surfaces have identical strikes (Fig. 1). Three-dimensional bedforms differ from two-dimensional bedforms in having one or more of the following characteristics: sinuous crestlines (either in plan form or in elevation), sinuous troughs (either in plan form or in elevation), or across-crest profiles that vary along the crestline. All of these variations produce three-dimensional cross-bedding—cross-bedding in which cross-bed strike varies within a set of cross-beds.

Invariable bedforms are defined here as bedforms that do not change in morphology or path of climb through time or space. Invariable bedforms deposit

sets of invariable cross-beds—sets in which all foresets, when considered in three dimensions, are geometrically identical. In contrast, variable bedforms change in morphology or path of climb through time or space; individual foresets in the sets of variable cross-beds are not geometrically identical.

Just as no bedforms are perfectly two-dimensional (because their crestlines cannot have infinite extent), no bedforms are perfectly invariable (because they cannot exist indefinitely without changing). Nevertheless, the terms "two-dimensional" and "invariable" are useful for describing bedforms with relatively simple morphology and behavior. No attempt is made here to define limits to the deviations from perfect two-dimensionality and perfect invariability that are allowable within these classes.

Bedforms can also be classified as transverse, oblique, and longitudinal, depending on their orientation relative to the long-term resultant sediment-transport direction. Transverse bedforms trend roughly parallel to the transport direction, longitudinal bedforms trend roughly normal to the transport direction, and oblique bedforms have intermediate trends. Previous studies have arbitrarily selected 15° as the maximum permissible divergence from perfectly transverse or perfectly longitudinal before bedforms are considered to be oblique (Hunter and others, 1983).

Recognition

Invariable two-dimensional bedforms and cross-bedding.— As a result of simple bedform morphology, all cross-beds and bounding surfaces generated by a set of identical two-dimensional bedforms have identical strikes (Fig. 1). This characteristic is visible in block diagrams, particularly in horizontal sections, and also in plots of cross-bed and bounding-surface dips (as illustrated in all computer-generated images in Figs. 1-29). This category of invariable two-dimensional structures includes most structures that have been called tabular sets of cross-beds. Not all tabular sets are two-dimensional, however; some sets with relatively planar set boundaries have cross-beds that are curved in plan form (Figs. 32, 33, and 44).

Cross-beds deposited by invariable two-dimensional bedforms are geometrically identical at all locations in the structure. Along-strike similarity of the bedding results from along-crest similarity of the bedforms, and down-dip similarity of the bedding results from the invariability through time of the bedform shape and behavior. Bounding surfaces scoured by invariable two-dimensional bedforms have the form of parallel planes, a characteristic which results from the migration in a constant direction of the parallel linear bedform troughs. Parallelism of bounding surfaces is recognizable in outcrops and in polar plots, on which the dips of bounding-surface planes plot as a single point (Figs. 1 and 5).

Variable two-dimensional bedforms and cross-bedding.— Variable bedforms change in morphology or path of climb while they migrate. Morphologic variability of two-dimensional bedforms is restricted to changes in height, spacing, asymmetry, or other parameters that determine across-crest profile. No other changes in morphology are possible without making the bedforms three-dimensional. Behavioral variability of two-dimensional bedforms is restricted to changes in the path of climb; changes in the path of climb can be caused by changes in the rate of deposition or changes in the rate of bedform migration.

Changing either the morphology or path of climb (or both) causes bounding surfaces scoured by two-dimensional bedforms to be curved instead of planar. The curved bounding surfaces produced by variable two-dimensional bedforms all have the same strike, whereas the curved bounding surfaces scoured by migrating three-dimensional bedforms vary in strike. These characteristics are recognizable in outcrop and in plots of bounding-surface dips; dips of cross-beds and bounding surfaces produced by variable two-dimensional bedforms plot along a single line through the center of the plot (computer images in Figs. 1 and 13-29).

Invariable three-dimensional bedforms and cross-bedding.— Invariable three-dimensional bedforms have a simpler behavior and more complex morphology than variable two-dimensional bedforms. The complex bedform morphology includes surfaces that dip toward a variety of directions; migration of these complex surfaces produces cross-beds that vary in direction of dip. These variations in dip direction can be seen in horizontal sections and in the dispersion of cross-bed dips in polar plots (Fig. 1). In contrast to the planar bounding surfaces scoured by invariable two-dimensional bedforms, the bounding surfaces scoured by three-dimensional bedforms are curved or trough-shaped (computer images in Figs. 1 and 32-79). Polar plots of the computer-generated bedding illustrate that the trends of the axes of such trough-shaped sets of cross-beds can be determined without direct observation of the axes. Trough-axis trends can be determined from random measurements of the poles of the bounding surfaces of the trough-shaped

sets; the trend of the trough axis is normal to the line along which the bounding-surface poles plot.

Variable three-dimensional bedforms and cross-bedding.— Three-dimensional bedforms can undergo a variety of changes that make the bedforms variable: morphologic changes such as fluctuations in height, asymmetry, or crestline sinuosity, or behavioral changes such as fluctuations in the speed or direction of migration of the main or superimposed bedforms. These changes cause the trough-shaped bounding surfaces scoured by the topographically low scour pits in the bedform troughs to become irregular. The irregularity can result from the up-and-down scour-pit migration caused by fluctuating depth of scour pits or from the back-and-forth migration of scour pits caused by fluctuating migration of the main bedforms or superimposed features.

Irregularity of the bounding surfaces is apparent in outcrop, particularly in horizontal sections, where paths of scour-pit migration are displayed most clearly. Irregularity of the bounding surfaces is also evident in plots of dips of bounding surfaces, because the dips plot as scatter diagrams (computer images in Figs. 1 and 58-79). In contrast, invariable two-dimensional cross-bedding has polar plots in which bounding-surface planes plot as a single point, two-dimensional variable cross-bedding has polar plots in which poles of bounding surfaces and cross-beds plot along the same line, and invariable three-dimensional cross-bedding has polar plots in which bounding surfaces plot as lines that are transverse or oblique to the direction of cross-bed dip.

Scale of Classification

Although the classification of cross-stratified deposits can be carried out at any scale, as the size of the sample increases, cross-bedding is less likely to be two-dimensional or invariable. For example, if the sample includes more than one set of cross-beds, then differences in cross-bed dip directions can be caused both by three-dimensionality of individual bedforms and by differences from one bedform to another. Similarly, if a bedform has plan-form sinuosities that are less regular than those of the computer-generated bedforms, then the different parts of a single bedform may deposit different kinds of structures. Including these different structures in a single sample could be expected to introduce scatter to the plots of bounding-surface poles (as do other causes of bedform variability). The computer simulations suggest, however, that the amount of scatter introduced by this cause of variability is relatively slight, apparently because the different parts of the bedform all migrate in the same direction (Figs. 36 and 45).

RELATIONS BETWEEN CROSS-BEDDING, BEDFORMS, AND FLOW

Approach

The usefulness of cross-bedding as a flow indicator results from the connections between cross-bedding, bedforms, and flow conditions. As a result of empirical and theoretical studies in the past few decades, it is now possible to predict crudely what bedform morphology results from flow conditions for many two-dimensional flows that are steady through time and uniform through space and for many two-dimensional oscillatory flows. Eventually, it may become possible to predict bedform morphology and behavior accurately for more complicated flows such as those that vary in strength or direction or the geologically more important flows that decelerate downcurrent (Rubin and Hunter, 1982). Ultimately, it may be possible to use cross-bedding to recreate current-meter-type records of paleocurrent directions and velocities. Even without quantitative fluid dynamics models, however, it is possible to infer bedform morphology and behavior from cross-bedding and to relate those interpreted characteristics qualitatively to flow conditions. The following section considers the controls of flow on bedform morphology and behavior and considers some of the general properties of the cross-bedding that is produced.

Plan-Form Geometry

Degree of three-dimensionality.— The extent of two- or three-dimensionality of cross-bedding is an important geometric property because three-dimensionality of bedding is an indicator of bedform three-dimensionality, and bedform three-dimensionality is an indicator of flow conditions. Interpreting flow properties from bedform three-dimensionality is complicated, because many processes influence the extent to which bedforms are two- or three-dimensional.

(1) Some kinds of bedforms, such as wind ripples, are inherently two-dimensional.

(2) Bedforms that are produced by reversing flows tend to be more two-dimensional than their unidirectional counterparts. For example, wave ripples are more two-dimensional than current ripples; sand waves in reversing tidal flows are more two-dimensional than sand waves or dunes in unidirectional flows such as in rivers; and linear eolian dunes,

which tend to form in reversing flows (Tsoar, 1983; Fryberger, 1979), are more two-dimensional than the barchanoid or crescentic dunes that form in unidirectional flows.

(3) Some workers have reported that the three-dimensionality of subaqueous bedforms increases with flow strength. According to some reports, current ripples are more three-dimensional at higher flow velocities—keeping mean depth constant—or at shallower depths—keeping mean velocity constant (Allen, 1968, 1977; Harms, 1969; Banks and Collinson, 1975). Middleton and Southard (1984, p. 7.59), however, disputed these findings and concluded that ''no definitive or unified picture of spacing, height, velocity, and plan geometry has emerged.'' There seems to be better agreement that large-scale subaqueous bedforms (dunes and sand waves) tend to be more three-dimensional at relatively high shear velocities or at relatively high velocities for any fixed depth (Allen, 1968; Southard, 1975).

(4) Ripples tend to be more two-dimensional where rapid deposition from suspension is occurring (Harms and others, 1982).

(5) Immature ripples have been reported to be more two-dimensional than more fully developed ripples (Ashley and others, 1982).

Although deposits of stoss-erosional two-dimensional bedforms are readily recognizable because their cross-beds dip toward the same direction (dispersion of dip directions is low), dispersion of cross-bed dip directions is not controlled entirely by bedform three-dimensionality; dispersion of dips is also influenced by bedform variability, behavior, and angle of climb. For example, reversals in the along-trough migration direction of lee-side scour pits increase the dispersion of cross-bed dips without changing bedform morphology or three-dimensionality (compare Figs. 38 and 59). Similarly, dispersion of cross-bed dips depends on the relative migration speeds of main bedforms and superimposed bedforms in situations where bedform morphology is constant (Fig. 46E and M). Because the dispersion of cross-bed dips depends on such factors as bedform variability and on the angle of climb (that is, dispersion is not determined uniquely by bedform morphology), much work remains to be done before three-dimensionality of bedforms can be quantitatively related to the dispersion of cross-bed dips.

Kinds of three-dimensionality.— It is obvious from examining bedforms in the field—and equally obvious when attempting to simulate bedforms mathematically—that there are at least two kinds of three-dimensionality: three-dimensionality caused by plan-form curvature and three-dimensionality caused by the superpositioning of positive or negative topographic features on bedforms that otherwise might be straight-crested. The effects of these different kinds of three-dimensionality on the geometry of cross-beds and bounding surfaces have not been adequately distinguished in previous studies. Regardless of the geometric details of the three-dimensionality, cross-beds deposited by three-dimensional bedforms vary in direction of dip, and traces of these cross-beds are curved in horizontal sections. Bounding-surface geometry depends, however, on the geometric details of the three-dimensionality. Where bedform troughs contain closed depressions (scour pits) such as those that occur between out-of-phase crests, lee-side spurs, or superimposed bedforms, the resulting bounding surfaces are shaped like troughs or truncated troughs, as illustrated in Figures 34 and 46. In contrast, bedforms with plan-form curvature but with troughs that do not vary in elevation produce bounding surfaces that are more nearly planar, as illustrated in Figure 32. Although the bounding surfaces produced by bedforms with scour pits thus differ considerably from those produced by bedforms lacking scour pits, distinguishing the deposits of bedforms with sinuous, linguoid, and lunate plan-form geometries is virtually impossible without exceptionally revealing horizontal sections (Figs. 32 and 34) or without unusually complete preservation of bedforms.

Despite considerable study, the hydraulic significance of specific plan-form shapes has not yet been quantitatively documented. Allen (1968) reported that ripples with in-phase crestlines form in weaker flows than ripples with out-of-phase crestlines, but as yet there is poor understanding of what flow conditions produce sine-shaped, linguoid, or lunate plan-form geometries or what flow conditions control the phase relations of bedforms with these different plan-form geometries.

The three-dimensionality of many bedforms results from superpositioning of bedforms or other topographic features rather than from bedform plan-form curvature. The superimposed topographic features include spurs and scour pits in bedform troughs, peaks and saddles on bedform crests, and small bedforms that may be superimposed at restricted or widespread locations on the main bedforms. Several experimental studies have found that lee-side spurs become more closely spaced with increasing flow strength (Allen, 1969, 1977; Banks and Collinson, 1975), but the results are difficult or impossible to apply to ancient bedforms, not merely because of disagreement about which is the proper measure of flow strength (Froude number or shear

stress) but because spacing of spurs has also been found to depend upon both flow strength and channel width (Allen, 1977).

Small bedforms are commonly superimposed on larger bedforms, and the migration directions of the two sets of bedforms often diverge. Two models have been proposed to explain bedform superpositioning: a fluctuating-flow model (Allen, 1978) and a multiple-boundary-layer model (Rubin and McCulloch, 1980). In the fluctuating-flow model, superimposed bedforms arise when flow conditions change and new bedforms are created before the old bedforms are destroyed. Superimposed bedforms that migrate in the same direction as the main ones are believed to indicate changes in flow strength (Allen, 1978), whereas superimposed bedforms that migrate in a different direction are believed to indicate changes in flow direction (Hereford, 1977; Elliott and Gardiner, 1981).

In the boundary-layer model, large bedforms create boundary layers (Smith and McLean, 1977) in which smaller bedforms can exist. The surface of the large bedform, like any sediment surface, is acted on by the overlying flow and is molded into a flat bed, ripples, dunes, or another bed configuration, depending upon the local flow conditions near the bed (Rubin and McCulloch, 1980). Superimposed bedforms formed in such steady flows are common in flumes (Guy and others, 1966), but most flume flows are so shallow that the resulting bedforms are small, and the superimposed dunes or sand waves, which are even smaller, are the size of ripples (Davies, 1982). In larger flumes, such as those that are on the order of a meter deep, large bedforms can be created, and the superimposed bedforms are large enough to be recognized as dunes or sand waves (Bohacs, 1981).

Many of the computer images in this publication illustrate depositional situations where two sets of bedforms simultaneously migrate in different directions. Although such behavior might seem unlikely, if not impossible, deposits produced by bedforms with this kind of behavior are common and can be readily explained by both fluctuating flow and multiple boundary layers. First, fluctuations in flow direction might alternately maintain two sets of bedforms. If the individual flow fluctuations transport small enough amounts of sediment relative to the sizes of the bedforms, then the two sets of bedforms will have the appearance of migrating simultaneously. Second, where the large bedforms are oblique to the flow direction, local flow on the lee side may take the form of a helix with an axis parallel to the bedform crestline (Allen, 1968). Bedforms created on the bed below such helical flow will develop in response to those local flow conditions and could be expected to have a different trend from the main bedforms. In addition to these processes that can maintain two sets of bedforms for long periods of time, two or more sets of bedforms can also exist temporarily at a site where flow conditions change and one set of bedforms is replaced by another (Fig. 79).

Invariable and Variable Bedforms and Cross-Bedding

In contrast to the dispersion in cross-bed dip directions that is caused by bedform three-dimensionality, bedform variability causes dispersion in inclination of bounding surfaces (Fig. 1). Where variable bedforms are two-dimensional, bounding-surface dips are dispersed in inclination but not in direction; where variable bedforms are three-dimensional, bounding-surface dips are dispersed in both inclination and direction.

Variability of bedforms arises from two kinds of processes: flow changes that cause bedforms to change in morphology or behavior, and bedform interactions that cause bedforms to change, even in steady flows. Processes that can cause variability even in steady flows include relatively random processes, such as splitting and merging of individual bedforms (Allen, 1973), and more systematic processes such as superpositioning of one set of bedforms on another (Rubin, 1987). Although distinguishing the deposits of variable and invariable bedforms is relatively simple, distinguishing variability produced by flow fluctuations from variability produced by superimposed bedforms is a difficult problem that has been the subject of many previous studies (McCabe and Jones, 1977; Hunter and Rubin, 1983; Terwindt and Brouwer, 1986; Rubin, 1987).

Fluctuating flow and superimposed bedforms can be expected to produce recognizably different kinds of structures because the effects of flow fluctuations are more widespread than the effects of bedform superpositioning. For example, changes in flow commonly cause entire trains of bedforms simultaneously to change in angle of climb (Figs. 13 and 14), or cause individual bedforms to change in profile for great along-crest distances. The foresets produced by such processes will extend for long distances along strike. In contrast, the effect of superimposed bedforms is more localized. For example, superimposed bedforms might not extend across the entire length of a main bedform or may arrive at different parts of the crestline of the main bedform at different times (Figs. 46, 65-67, and 72-74). These more localized processes deposit foresets with more limited along-crest extent, and the foresets deposited by

superimposed bedforms commonly dip in a different direction from the bounding surfaces, as illustrated in many of the computer simulations.

Flow fluctuations can be either random or cyclic. Cyclic flow fluctuations can produce cyclic foresets by causing cyclic fluctuations in bedform size (Figs. 15-17), cyclic fluctuations in bedform asymmetry or migration speed (Figs. 18-24, 29, 58, 67, and 77), or cyclic avalanching processes (Hunter, 1985). Any of these fluctuations in flow can produce annual cycles of eolian foresets (Stokes, 1964; Hunter and Rubin, 1983) and can also produce tidal cross-bedding with a double cyclicity (neap-spring and ebb-flood), as described by Boersma (1969) and Terwindt (1981). Cyclic cross-bedding can be produced even in steady flows by superimposed bedforms that transport sediment in cyclic pulses across the main bedform crest or along the lee slope (McCabe and Jones, 1977; Hunter and Rubin, 1983; Rubin, 1987).

Some of the computer-generated bedforms are perfectly straight-crested and have superimposed bedforms that exactly parallel the main bedforms (Figs. 25 and 27); such bedform assemblages produce structures that are virtually indistinguishable from those produced by fluctuating flow. In the real world—or in more realistic simulations, such as those shown in Figures 65 and 66—superimposed bedforms do not exactly parallel the main bedforms for long distances along-crest, and the deposits of superimposed bedforms are more readily recognized. In real deposits, the distinction between fluctuating-flow compound cross-bedding and superimposed-bedform compound cross-bedding can also be based on nongeometric characteristics of the bedding. For example, fluctuating-flow cross-bedding may contain mud drapes, indicating sediment fallout during intervals of low-velocity flow. Similarly, reversals in migration direction of superimposed bedforms are direct indicators of flow reversals.

Transport Direction and Bedform Orientation

Transport direction inferred from cross-bed dips.— One of the most important applications of cross-bedding analysis is the determination of paleocurrent directions. The traditional approach has been to measure large numbers of cross-bed dips and to presume that the mean cross-bed dip direction represents the paleocurrent or paleotransport direction. This approach is probably quite reliable for the deposits of transverse bedforms and for many three-dimensional longitudinal bedforms. It is also quite likely that if a deposit was produced by a diverse assortment of bedforms or by bedforms with three-dimensional superimposed bedforms, then the divergence between the paleocurrent direction and the cross-bed dip direction in individual beds will, in many cases, merely cause scatter to the data, rather than introducing a systematic bias. Oblique bedforms, however, can deposit cross-beds that dip with a systematic divergence from the direction of sediment transport (Figs. 42-46 and 69-74).

Determining the paleotransport direction solves only half the problem of distinguishing the deposits of transverse, oblique, and longitudinal bedforms; bedform orientation must also be determined. The most straightforward technique for determining bedform orientation is by inspecting sections that parallel the generalized depositional surface. Such sections contain indications of bedform orientation, such as aligned "fingertip structures" (Figs. 38 and 46) or foresets that extend laterally for distances that are large relative to the the bedform spacing. Distinguishing the deposits of transverse, oblique, and longitudinal bedforms is useful in determining fluctuations in flow direction, because some kinds of bedforms such as longitudinal dunes tend to form in reversing flows. Moreover, recognition of the different kinds of bedforms is the first step toward understanding how bedform alignment is controlled in directionally varying flows.

Transport direction inferred from trends of trough axes.— Trends of trough axes are commonly used to infer paleotransport directions. As illustrated by many of the computer images, trough axes have the same trend as the displacement direction of the bedform surface, but the displacement direction of the bedform surface does not necessarily parallel the transport direction. The divergence between the trough-axis trend and the resultant bedform transport direction is most pronounced where a trough-shaped set is produced by migration of a scour pit that is bounded on one side by the lee slope of the main bedform and on the adjacent sides by much smaller lee-side spurs or superimposed bedforms. In such a situation, a unit distance of scour-pit migration in an along-trough direction represents less transport than an equal distance of transport in a direction normal to the main bedform because of the difference in size of the bedforms migrating in bedform-normal and bedform-parallel directions. The following section considers this problem in detail.

Transport direction inferred from along-crest and across-crest transport.— Bedforms can be classified as transverse, oblique, or longitudinal by their orientation relative to the long-term resultant sediment-transport direction. Transverse bedforms trend roughly normal to the transport direction, longitudinal bedforms trend roughly parallel to the transport direction, and oblique bedforms have intermediate trends.

The usual approach to classifying bedforms using this scheme has been to use current-velocity measurements and transport-rate equations to calculate the transport direction and then to measure the deviation between the calculated transport direction with the bedform trend. An alternative is to use the sediment transport represented by bedform migration to determine the relative rates of across-crest and along-crest sediment transport. The use of bedform height, shape, and migration speed to determine the rate of sediment transport is well known, having been proposed in 1894 (work by Deacon referenced in Goncharov, 1929; Hubbell, 1964), applied in fluvial studies in 1955 (Benedict and others), tested in flumes in 1965 (Simons and others), and applied to the study of paleotidal flow velocities from tidal-bundle thicknesses in the 1980s (Allen, 1981; Allen and Homewood, 1984). The rate of sediment transport represented by bedform migration (called the bedform transport rate by Rubin and Hunter, 1982) is given by

$$i = VHk \qquad (1)$$

where i is the bedform transport rate (expressed in units of bulk volume per unit time per unit width), V is the rate of bedform migration, H is bedform height, and k is a dimensionless shape factor equal to A/HL; A is bedform cross-sectional area (measured in a vertical plane parallel to the transport direction), and L is bedform spacing. Bedforms that are triangular in profile have a shape factor (k) equal to 1/2. Equation (1) is correct only if bedforms are transverse to the transport direction or if i, V, H, and k are measured in a plane that parallels the transport direction. Where bedforms are not transverse and where i, V, H, and k are measured normal to the bedform trend, equation (1) must be modified to

$$i = \frac{VHk}{\sin\alpha} \qquad (2)$$

where α is the angle between the bedform trend and the resultant transport direction (90° for a transverse bedform and 0° for a longitudinal bedform). Even where V, H, and k are known, equation (2) cannot be used to solve for the orientation of the transport direction relative to the trend of a two-dimensional bedform, because an infinite number of transport vectors can produce a given migration rate; a small vector normal to the bedform crestline can produce the same migration rate as a larger vector that more nearly parallels the bedform trend.

Although equation (2) cannot be used to solve for α where bedforms are perfectly two-dimensional, that equation can be modified for such use where bedforms are three-dimensional. Conceptually, the approach is to determine the unique transport vector that simultaneously would cause the observed migration of two sets of bedforms. Algebraically, this is accomplished by solving equation (2) simultaneously for the transport represented by two sets of bedforms. The solution is given by

$$\alpha_1 = \tan^{-1}\left[\frac{V_1 H_1 \sin\beta}{V_1 H_1 \cos\beta - V_2 H_2}\right] \qquad (3)$$

where the subscripts refer to the two sets of bedforms, and β is the angular divergence of the migration directions of the two sets of bedforms.

Equation (3) can also be applied to a single set of bedforms, if they are three-dimensional. In such a situation, β is equal to 90°, V_2 is the along-crest migration speed of the plan-form sinuosities, and H_2 is the mean height of the bedforms measured along profiles parallel to the generalized trend of the bedforms. In the computer-generated depositional situations, H_2 was measured from contour maps of the bedform topography. Although equation (3) cannot be used with perfectly two-dimensional computer-generated bedforms, most real bedforms, including many that would be considered two-dimensional, are probably three-dimensional enough to use this approach.

Note that the transport direction given by equation (3) considers only the fraction of transport that is represented by bedform migration. That transport direction will parallel the direction of total transport only if the bedforms are equally effective traps for sediment transported in different directions across their surfaces. This property is less likely to be met where the two sets of topographic features have grossly different morphology. For example, plan-form sinuosities may be less effective traps for sediment transported along-crest than are the main lee slopes for sediment transported across-crest.

A second difficulty can arise when using equation (3) to determine the transport direction in those situations where transport is represented by more than two sets of topographic features: different pairs of features give different calculated transport directions. Such a discrepancy occurred in several of the situations modeled in this publication; in those cases, preference was given to transport directions calculated with respect to pairs of bedforms with similar morphology. Despite these limitations, the technique represented by equation (3) is a useful approach for determining the transport direction from bedform migration.

The relative heights and migration speeds of main bedforms and superimposed bedforms can rarely be determined from cross-bedding, and along-crest and across-crest components of transport must be treated qualitatively rather than quantitatively. The transport direction represented by the migration of perfectly two-dimensional bedforms cannot be determined more accurately than approximately 180° (within 90° of the bedform migration direction); transport toward any direction within this range will cause lateral migration of the bedforms. In contrast, where superimposed bedforms are present and are migrating toward a direction different from that of the main bedform, the transport direction can be limited to a single quadrant (limited to one hemisphere by the migration of the main bedforms and limited to one-half of that hemisphere by the right-hand or left-hand migration of the superimposed bedforms). Along-crest migration of superimposed bedforms is recognizable merely by inspection of some outcrops (particularly in sections that parallel the generalized depositional surface) and by an asymmetric distribution of cross-bed planes relative to bounding-surface planes (computer images in Figs. 42-46 and 69-74).

Controls of bedform alignment.— Compared to the numerous studies of equilibrium bedform size and shape, the study of how bedform orientation varies as a function of flow conditions has received surprisingly little work. With the exception of longitudinal eolian dunes, many workers believe that ripples, dunes, and sand waves are inherently transverse bedforms. All of these kinds of bedforms, however, can be oblique to the resultant transport direction. Obliquity can result from nonuniform flow conditions that cause one end of a bedform crestline to outrun the other end (Dietrich and Smith, 1984) or from nonuniform conditions that cause the transport direction to rotate downcurrent over a distance that is too short for the bedform to respond (Rubin and Hunter, 1985).

Oblique bedforms can originate even in uniform flows. Experiments conducted on a rotatable sand-covered board (Rubin and Hunter, 1987) have shown that transverse, oblique, and longitudinal wind ripples can be created in bidirectional winds merely by varying two parameters: the angle between the two winds (the divergence angle) and the proportions of sand transport in the two directions (the transport ratio). Transverse bedforms were created when the divergence angle was less than 90°, when the transport ratio was large, or when the divergence angle approached 180° and the transport ratio was not equal to unity; longitudinal bedforms were created when the divergence angle was greater than 90° and the transport ratio approached unity; and oblique bedforms formed when the divergence angle was greater than 90° and the transport ratio was between unity and approximately eight (Appendix B, Fig. B-1). In all experiments, the bedforms followed the rule of maximum gross bedform-normal transport: the bedforms had the trend that was subject to the maximum gross (transports in opposite directions are summed as two positive numbers) transport across bedforms.

In these experiments, the fact that some bedforms were longitudinal (parallel to the resultant transport direction) was merely coincidental. Such bedforms are more properly thought of as bedforms that are as transverse as possible to the separate transport vectors. These experimental results are compatible with previous field studies in which it was shown that longitudinal dunes can form without flow parallel to their crestlines (Tsoar, 1983) and that linear dunes tend to form in reversing wind regimes (Fryberger, 1979; Twidale, 1981).

Flow Velocities

Several techniques have been used to estimate paleoflow velocities. One technique consists of identifying the kind of bedform that produced a deposit and then researching the flow conditions that produce such bedforms. Empirical data relating bedform morphology and behavior to flow conditions are available for wide ranges of conditions and bedforms (Guy and others, 1966; Southard, 1971; Middleton and Southard, 1984) and for specific kinds of bedforms such as current ripples (Harms, 1969; Banks and Collinson, 1975; Allen, 1977; Ashley and others, 1982; Middleton and Southard, 1984), wave ripples (Bagnold, 1946; Inman, 1957; Komar, 1973, 1974; Dingler, 1974; Clifton, 1976; Allen, 1979), wind ripples (Bagnold, 1941; Sharp, 1963; Walker, 1981), subaqueous dunes and sand waves (Stein, 1965; Dalrymple and others, 1978; Rubin and McCulloch, 1980; Costello and Southard, 1981; Middleton and Southard, 1984), and antidunes (Gilbert, 1914; Kennedy, 1969; Hand, 1974). Dimensional analysis has enabled empirical bed-phase relations to be extended to flows with unusual sediments, unusual fluids, or unusual temperatures (Southard, 1971), but other bed phases (such as marine and estuarine mud waves, furrows, adhesion ripples, and eolian dunes) have been less well quantified.

In rare cases paleoflow velocities can also be estimated by using bedform heights and migration speeds to determine paleotransport rates using equation (2). Transport rates can then be converted to flow velocities using empirical or theoretical transport-rate relations. This technique can only be applied to deposits where the bedform migration speeds can be determined; application of the technique also requires knowing the bedform height and the bedform trend relative to the transport direction. In the few cases where this technique has been applied, bedforms have been presumed to be transverse, and migration speeds have been determined from the distance of bedform migration caused by currents of known duration such as tidal currents (Allen, 1981; Allen and Homewood, 1984) or annual wind cycles (Hunter and Rubin, 1983).

SELECTION OF ILLUSTRATIONS

The computer model used to create the images in this publication uses 75 input parameters to describe bedform morphology and behavior. To illustrate three values for each variable, in combination with three of every other variable, would require more than 10^{35} images; and the effect of varying some of the parameters would still be insufficiently illustrated. The computer images in this publication are not a random sample of the structures of such a thorough set of images. Some images were selected to illustrate the kinds of detailed cross-bedding interpretations that are possible; some illustrate useful interpretive techniques such as distinguishing transverse, oblique, and longitudinal bedforms; some demonstrate that grossly different structures are indistinguishable in some outcrop planes; some simulate existing structures and demonstrate that the behavior of bedforms must be different from what is commonly expected; some demonstrate the use of computer graphics as a research tool (by showing the results of repeated trial-and-error computer experiments that duplicate real examples of cross-bedding). Computer images that illustrate more complicated and random bedform behavior and morphology were generally not included in this compilation, because such images are so complicated that they are nearly as incomprehensible as the bedding that they simulate.

In most of the computer simulations the angle of climb was adjusted so that the upper half of each bedform was eroded and the lower half was preserved. This angle of climb is probably higher than usual for most depositional situations and higher than is even possible in many situations (Rubin and Hunter, 1982). Although a lower angle of climb would probably be a more accurate approximation of the average depositional situation, using a larger angle of climb has the benefit of producing thicker beds (and more clearly displayed structures) without detracting from the results.

Most of the depositional situations are illustrated with more than one computer image. All of the situations include at least one image that shows bedform morphology and internal structures; several of the situations include a second image that shows the change in morphology through time or that shows enlarged details of the bedding. Three-dimensional structures (and a few examples of two-dimensional structures) are also illustrated with block diagrams that have horizontal sections at the top of the block. The blocks are oriented with their sides parallel and normal to the crestlines of the main bedforms. Most of the three-dimensional structures are illustrated with a second block diagram in which the structure is rotated to give vertical sections at additional orientations. Where the main bedforms are perfectly transverse or perfectly longitudinal, the structures in the second block diagram are rotated by $30°$; the two vertical sections in this block, together with the two in the block with sides parallel and normal to the bedform crestlines, display vertical sections that are oriented at angles of $0°$, $30°$, $60°$, and $90°$ to the crestlines of the main bedforms. Where the bedforms are oblique to the transport direction, the structure in the second block diagram is rotated so that the sides of the block are parallel and normal to trough axes. Most of the depositional situations also are illustrated with plots of directional data (migration directions of bedforms and scour pits, the bedform transport direction, and dips of cross-beds and bounding surfaces).

The field photographs that are included were also selected for their instructive value. The photographs generally accompany a computer image that illustrates a similar aspect of deposition; however, the field situations commonly differ in other aspects from the computer images. For example, the field example chosen to illustrate cyclic variations in the angle of climb (Fig. 14, analogous to the computer image in Fig. 13) was actually deposited on the lee slope of a larger topographic feature—a property not modeled in the computer image. In other words, the field examples illustrate key aspects of the computer images but do not necessarily duplicate all aspects of the computer-generated situations.

APPLICATION OF RESULTS

The images and interpretive techniques in this publication can be used for several purposes: visualizing how migrating bedforms deposit complicated cross-stratified beds, predicting the internal structure of bedforms that have a known morphology and behavior, interpreting the behavior of bedforms from their morphology and internal structure, or

reconstructing the morphology and behavior of bedforms that deposited cross-stratified beds that are observed in the field (either to learn about the bedforms that produced a specific deposit or to learn about how bedforms behave in general). For convenience all of the computer-generated depositional situations are cross-referenced in Appendix C.

Unfortunately for the field geologist, interpreting cross-bedding is often more complicated than matching illustrations and outcrops, because the curvature and orientation of outcrops can cause radical variations in the appearance of a single structure. Moreover, different bedform assemblages can produce structures that are so similar in single vertical sections that the structures are virtually indistinguishable. A more powerful investigative technique is to measure cross-bed strikes and dips and make a three-dimensional map of the beds that are being studied. These observations can then be compared with the computer-generated examples. For even more precise bedform reconstructions, observed structures can be reproduced by trial-and-error computer simulation (Rubin, 1987).

UNIQUENESS OF SOLUTIONS

The usefulness of the following computer-generated cross-bedding images depends on the uniqueness of the results. Specifically, can more than one structure result from a specific depositional situation, or can more than one depositional situation produce similar structures? The answer to the first question is that only one structure can result from a given depositional situation, if the bedform morphology and behavior are described in sufficient detail. If the situation is merely described qualitatively, many structures can result. For example, Figure 46 illustrates 14 different structures produced by main bedforms with superimposed bedforms migrating toward a diverging direction. In other words, no single image can represent all structures formed by a particular depositional process.

More serious interpretive problems arise if two or more depositional situations can produce the same structure; interpretation of real bedding structures then becomes equivocal. Such is certainly the case if the bedding structures are not observed in three dimensions. As illustrated in the following section, grossly different processes can produce structures that are virtually indistinguishable in a single two-dimensional exposure. In contrast, the computer experiments illustrated in the following figures suggest that structures that are similar in three dimensions are rarely produced by different processes.

COMPUTER IMAGES

Invariable Two-Dimensional Bedforms and Cross-Bedding

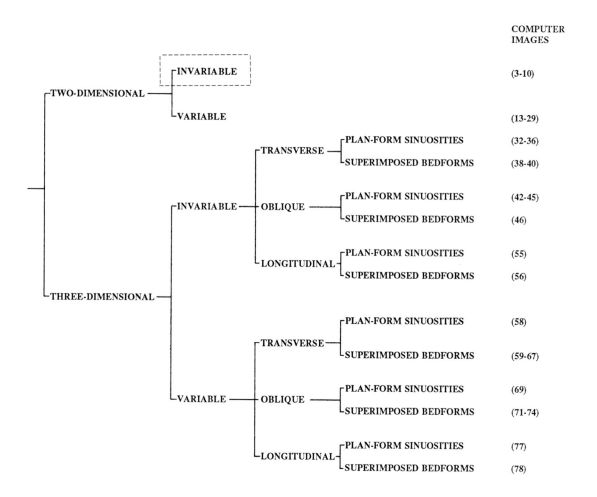

FIG. 2.— Schematic diagram showing the sequence in which illustrations are presented. Dashed line shows which structures are included in the following section.

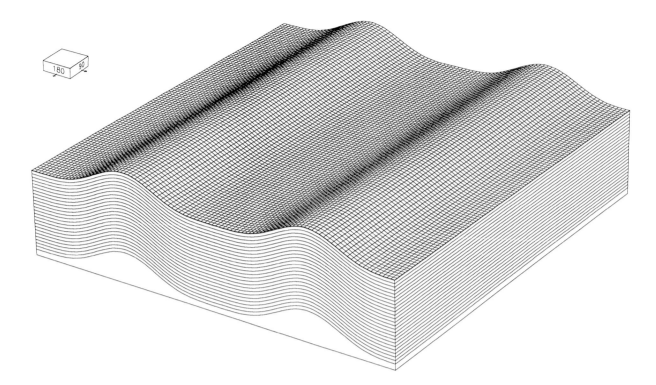

FIG. 3.— Structure formed by two-dimensional bedforms climbing vertically. Other names for similar structures are: ripple-laminae in phase (McKee, 1939, 1965), sinusoidal lamination (Jopling and Walker, 1968), draped lamination (Gustavson and others, 1975; Ashley and others, 1982), and complete rippleform lamination (Hunter, 1977). Most of these earlier terms are less restrictive as they also apply to structures deposited by bedforms climbing at nonvertical stoss-depositional angles and to structures deposited by three-dimensional bedforms. Vertically climbing bedforms that are three-dimensional or variable (or both) are simulated in Figures 18, 19, 55, 56, 77, and 78. Real structures deposited by bedforms that at times climbed vertically are shown in Figures 7, 20, and 64.

RECOGNITION: This structure differs in origin from other invariable two-dimensional cross-bedded structures by having a higher (vertical) angle of climb. The vertical angle of climb is indicated by the vertical alignment of bedform crests and troughs. Interpretation of bedform morphology is trivial, because the bedforms climb at such a high angle that depositional surfaces are completely preserved. The origin of the structure can also be recognized from the plots of cross-bed and bounding-surface dips. Stoss-depositional climb is indicated by an absence of bounding surfaces, and a symmetrical bedform shape is suggested by symmetrical dip patterns relative to the center of the plot. The plot that is shown includes the dips of cross-beds in vertical profiles at many locations on the bedform surface. If a separate plot were shown for each vertical profile, dips of all beds in each profile would plot as a single point, because the dip azimuth and inclination in each profile are constant with depth.

ORIGIN: This structure is relatively rare because vertical climb of bedforms requires unusual circumstances: a flow that maintains bedforms and transports sediment to the depositional site but does not cause the bedforms to migrate. These conditions are met or approximated:

(1) in flows over nonmigrating transverse bedforms such as some antidunes;

(2) in steady flows over perfectly aligned longitudinal bedforms;

(3) in directionally varying flows where bedforms trend exactly parallel to the resultant sediment transport direction (that is, the bedforms are perfectly longitudinal), and where short-term fluctuations in the transport direction move so little sediment that the bedforms do not migrate laterally or change shape with the individual fluctuations; unsteady flows that can produce longitudinal bedforms are those in which the transport direction varies by an angle of between 90° and 180° and in which transport from the two directions is equal (Fig. B-1; Rubin and Hunter, 1987);

(4) in unsteady flows where bedforms cease to be active, and vertical fallout drapes the bedforms with

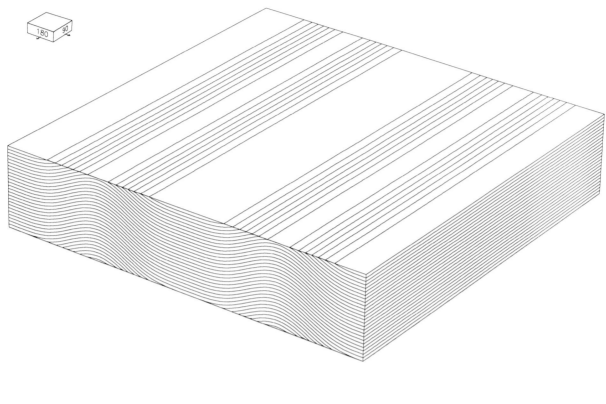

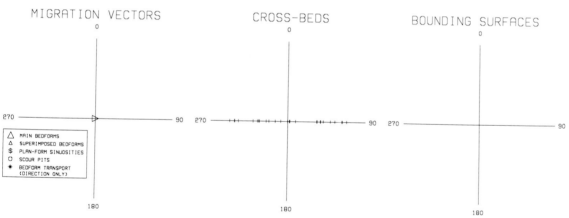

layers of sediment; this is probably one of the more common processes for creating thin deposits of vertically climbing bedforms, but, if the flow does not actively maintain the bedforms, vertical fallout will reduce the bedform height and eventually will destroy or bury the bedforms;

(5) in flows where slow-velocity sediment-laden flows rain sediment down preferentially on the upcurrent-facing slopes of bedforms, thus balancing the volume of sediment trapped on lee slopes. Under some flow conditions, the rate of deposition on upcurrent-facing slopes can exceed that on lee slopes, and bedforms migrate upcurrent. This process may explain the

commonly observed upslope migration of deep-sea mud waves, and this process seems to be the best explanation for the slight upcurrent migration exhibited by some ripples climbing at nearly vertical angles, such as those illustrated by Jopling and Walker (1968, center of fig. 7). Sediment cohesion may limit stoss-side erosion, thereby enabling higher angles of climb or upcurrent migration (Jopling and Walker, 1968).

Because this structure can be produced by transverse and longitudinal bedforms, and probably also by oblique bedforms, it is not a useful indicator of transport directions.

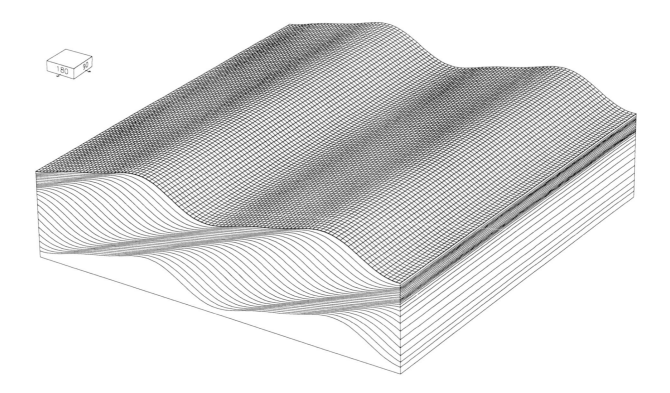

FIG. 4.— Structure formed by two-dimensional, stoss-depositional bedforms climbing at a subvertical angle. This structure is a two-dimensional example of the more general class of structures that have been called type 3 ripple-drift cross-lamination (Walker, 1963), type B and C ripple-drift cross-lamination (Jopling and Walker, 1968), supercritical climbing-ripple structure (Hunter, 1977), depositional-stoss climbing-ripple cross-stratification (Harms and others, 1982), and stoss-depositional climbing-ripple structure (Rubin and Hunter, 1982).

RECOGNITION: Like depositional surfaces in structures deposited by vertically climbing bedforms, depositional surfaces in this structure are not truncated. Bedform morphology and behavior are readily inferred from the completely preserved depositional surfaces.

ORIGIN: The conditions required to produce subvertical climb are not quite as unusual as the conditions necessary to produce vertical climb, because subvertically climbing bedforms do not remain stationary while sediment is transported to the depositional site. Even subvertical stoss-depositional climb requires somewhat unusual conditions, however. Net deposition on stoss slopes requires a rate of deposition that approaches or exceeds the rate of bedform migration. These conditions can be met either by a relatively rapid rate of deposition or by a relatively low rate of bedform migration. Rapid rates of deposition imply high transport rates and fallout from suspension, as explained by Ashley and others (1982), whereas slow rates of bedform migration imply low transport rates. An ideal situation for meeting these conflicting requirements is in a flow that undergoes a downcurrent decrease in transport rate, such as would occur where the near-bed velocity decreases. In such a situation, sediment can be transported to the depositional site at a rapid rate while bedforms migrate slowly. The conditions favoring this rapid deposition from suspension occur in fluvial flows (Fig. 7) and turbidity currents, and many excellent examples of this structure occur in deposits of such flows.

Low rates of bedform migration can also occur where bedforms are poor traps for the sediment that is being transported (as is the case with antidunes) or where bedforms trend nearly parallel to the transport direction. Although a longitudinal bedform orientation can reduce the bedform migration speed and thereby increase the angle of climb (equation 2 and Rubin and Hunter, 1985), the alignment with the flow must be exceptionally exact before bedforms can climb at stoss-depositional angles without deposition from suspension (Rubin and Hunter, 1985).

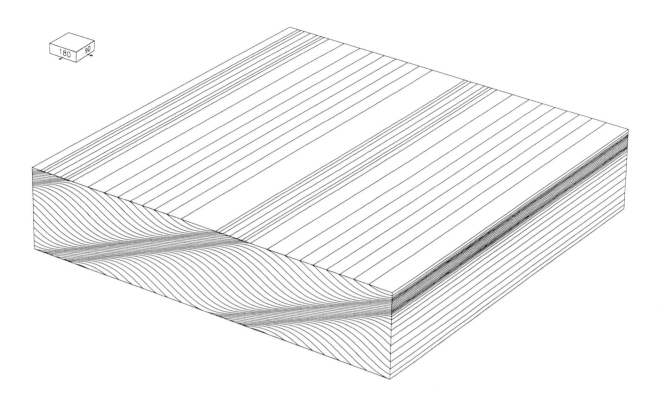

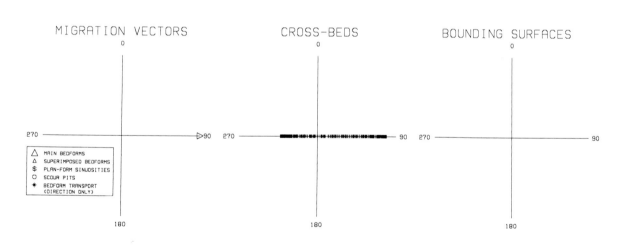

MIGRATION VECTORS

CROSS-BEDS

BOUNDING SURFACES

△ MAIN BEDFORMS
△ SUPERIMPOSED BEDFORMS
$ PLAN-FORM SINUOSITIES
○ SCOUR PITS
* BEDFORM TRANSPORT
  (DIRECTION ONLY)

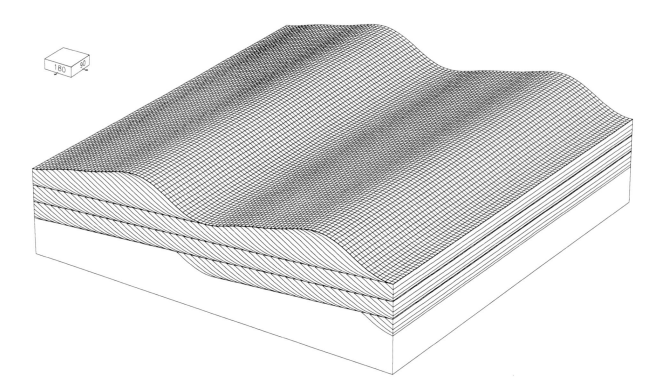

FIG. 5.— Structure deposited by two-dimensional bedforms climbing at a stoss-erosional, lee-depositional, net-positive angle of climb. Other (more general) names for this structure are ripple drift (Sorby, 1859), type 1 ripple-drift cross-lamination (Walker, 1963), climbing-ripple structure (McKee, 1965), climbing-ripple cross-lamination (Allen, 1972), subcritically climbing translent stratification (Hunter, 1977), erosional-stoss climbing-ripple cross-stratification (Harms and others, 1982), and stoss-erosional climbing-ripple structure (Rubin and Hunter, 1982). Most of the computer images in this volume illustrate stoss-depositional climb of bedforms, but with more complex behavior or morphology than shown here. Real examples of relatively simple stoss-erosional cross-stratification are shown in Figures 6 and 7.

RECOGNITION: Structures deposited by bedforms climbing at stoss-erosional angles lack complete rippleform laminae and instead have erosional bounding surfaces that separate the sets of cross-beds deposited by individual bedforms. This lack of completely preserved bedforms makes it difficult or impossible to determine bedform height and spacing.

Cosets of cross-beds can originate by several processes other than bedform climbing: superpositioning of delta-like sediment bodies (Jopling, 1965), migration of potholes (Hemingway and Clark, 1963), migration and independent planing off of bedforms (Stride, 1965), and buildup and deep truncation of

bedforms (Stokes, 1968). An origin by bedform climbing can be demonstrated by (1) an angular relation between climbing translent strata and more nearly isochronous underlying or overlying strata, (2) a change in the path of bedform climb within a coset of cross-strata, the change having taken place simultaneously across the bedform field (Figs. 7, 13, and 14), or (3) recognition of distinctive foresets that were deposited simultaneously on the lee slopes of adjacent bedforms (Fig. 6). Additional examples of climbing translent strata that are recognizable using these criteria are illustrated by Rubin and Hunter (1982).

Because this structure does not contain preserved bedform crests, except in rare cases at the top of the coset, the crestline trend is usually determined from the cross-bed strike. This technique is based on an implicit assumption that the bedforms that deposited the cross-beds were migrating over a horizontal depositional surface. A technique that is not restricted to horizontal depositional surfaces is to determine the trend of the line of intersection of cross-bed planes and bounding-surface planes. This can be accomplished by inspection at the outcrop or by the use of a stereonet. This technique is useful for determining the trend of small bedforms superimposed on the inclined lee surfaces of larger bedforms, as discussed in detail in Figure 46.

Unlike the stoss-depositional bedforms in Figures 3 and 4, the bedforms in this example undergo

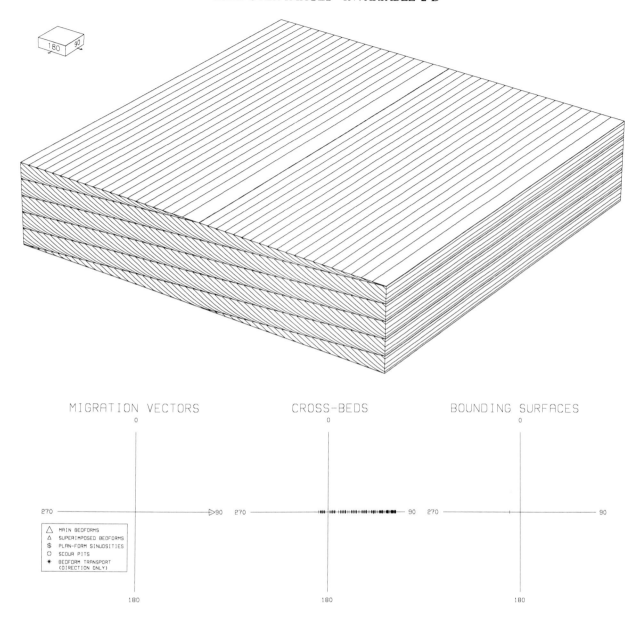

MIGRATION VECTORS

CROSS-BEDS

BOUNDING SURFACES

△ MAIN BEDFORMS
△ SUPERIMPOSED BEDFORMS
$ PLAN-FORM SINUOSITIES
○ SCOUR PITS
✳ BEDFORM TRANSPORT
    (DIRECTION ONLY)

erosion on their stoss sides, thereby producing upcurrent-dipping bounding surfaces. The bounding surfaces scoured by the migrating bedform troughs dip upcurrent, because deposition causes the elevation of the bedform trough to climb downcurrent. Consequently, the bounding surfaces dip in a direction opposite to that of most of the foresets; the lower-most foresets dip upcurrent because they are tangent to the bounding surfaces. A vertical profile of dips through this structure would show steep downcurrent dips at the top of each set of cross-beds, a reduction in inclination downward through the set, low-angle upcurrent dips in beds immediately overlying the lower bounding surface of the set, and an abrupt change to steep downcurrent dips when the bounding surface is crossed and the underlying set is sampled.

ORIGIN: In most flows, rates of deposition are much less than rates of bedform migration, and the resulting angles of bedform climb are generally so small that where bedforms climb, they usually climb at stoss-erosional angles. Consequently, of the structures deposited by migrating bedforms, those produced by stoss-erosional, lee-depositional angles of climb are the most common. Although some structures deposited by climbing wind ripples (Fig. 6), oscillation ripples, and ripples in flows with high rates of deposition from suspension are almost as perfect and regular as those in this example, bedforms in most other flows tend to be more three-dimensional, or they tend to reverse direction of migration, thereby changing the path of climb.

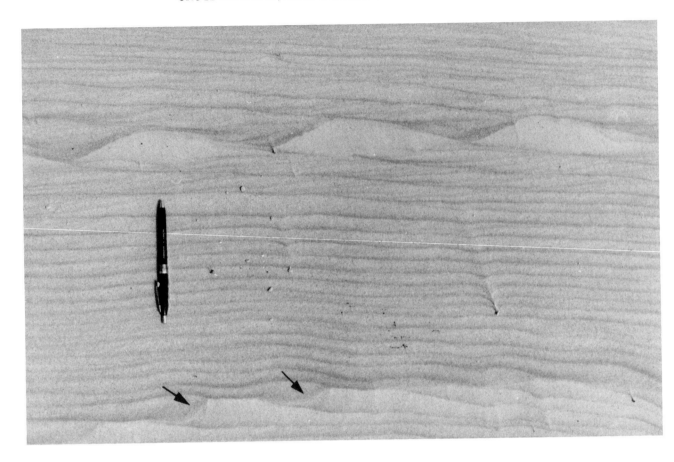

FIG. 6.— Structure deposited by two-dimensional wind ripples climbing at a stoss-erosional, lee-depositional, net-positive angle of climb; modern eolian deposits from Padre Island, Texas, photographed by Ralph Hunter. This structure is a real example of the structure generated by computer in Figure 5.

RECOGNITION: This structure was deposited by wind ripples that migrated from right to left. The path of ripple climb was steady from near the bottom of the photograph to immediately below the buried ripples. A temporary change in flow conditions, possibly a change in the flow direction, caused the ripples to be preserved at one horizon.

As the ripples migrated, they left behind cross-laminated beds. The beds have a mean thickness that is equal to approximately half the ripple height, which indicates that during the time that each ripple migrated one wavelength, deposition raised the bed elevation by approximately half the ripple height.

In the plane of the photograph, the bounding surfaces that separate the cross-laminated beds dip toward the right or are relatively horizontal, whereas the generalized depositional surface (the bed surface if the ripples were smoothed off) dips at a low angle toward the left. The inclination of the depositional surface can be recognized by connecting the troughs or crests of the preserved ripple forms. The inclination of the depositional surface also can be recognized by connecting the distinctive dark-colored laminae that were deposited simultaneously on the lee sides of adjacent ripples (indicated by arrows). The change in sand color from light to dark occurred at approximately the same time that the ripples changed their path of climb, suggesting that a change in wind direction or wind strength simultaneously introduced sediment from a new source and changed the ripple morphology or behavior.

FIG. 7.— Structure formed by ripples climbing at stoss-erosional, stoss-depositional, and vertical angles; fluvial deposits of the Colorado River, Grand Canyon National Park, Arizona. The analogous computer-generated structures are shown in Figures 3-5. Note pencil for scale.

RECOGNITION: In this example the angle of climb was at times stoss-erosional (E), stoss-depositional (D), and vertical or slightly upcurrent (V). This systematic change in ripple behavior indicates a change in flow conditions, but other variations in the structure are more random and therefore probably represent differences in behavior or morphology of individual ripples. For example, one ripple apparently decayed while migrating; that ripple migrated from the lower right side toward the center of the photograph, at which point the ripple height decreased so much that the ripple almost disappeared. In the overlying deposits (where the ripples climb vertically), a new ripple appeared in the same part of the ripple train. From this single vertical section it is impossible to determine whether these appearances and disappearances are real changes through time in the ripple train or whether they are merely apparent changes caused by migration of three-dimensional bedforms obliquely through the outcrop plane.

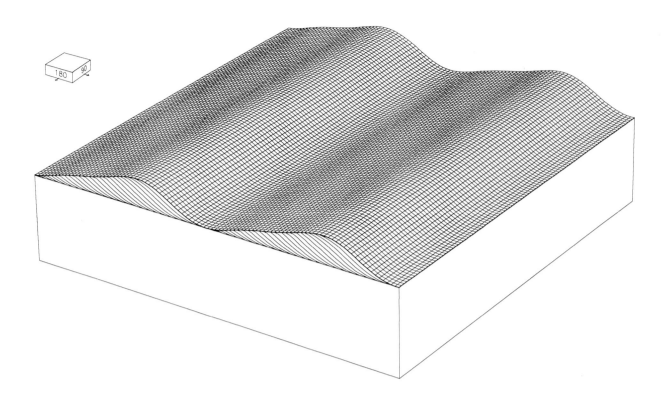

FIG. 8.— Structure formed by two-dimensional bedforms migrating without deposition (climbing at an angle of 0°).

RECOGNITION: This structure is bounded on its base by a plane that is tangent to the bedform troughs; the plane results from the passage of all bedform troughs along a single plane. In the pure form that is illustrated here, this is not a useful structure for interpreting bedform morphology or behavior; unless the bedforms become buried, the only feature that is preservable is the plane along which the bedforms migrated. If flow conditions change, however, the bedforms may climb at a positive angle, in which case the plane scoured by climb at an angle of 0° becomes a useful indicator of the generalized depositional surface (Figs. 13 and 14). When this surface has been identified, the angle of climb and bedform spacing can be determined (Rubin and Hunter, 1982, 1984). Cross-bed and bounding-surface dips (not shown) are similar to those of the positively climbing bedforms in Figure 5, except that in this example the bounding surfaces are horizontal, and foresets have no upcurrent dips.

ORIGIN: The flow processes that produce this structure—bedform migration with neither deposition nor erosion—are probably extremely common in nature, but, as these processes are nondepositional, the structure is almost certainly under-represented in the geologic record. Approximations of this structure, however, can be preserved where bedforms climb at extremely small positive angles, thereby depositing thin laminae that nearly parallel the generalized depositional surface.

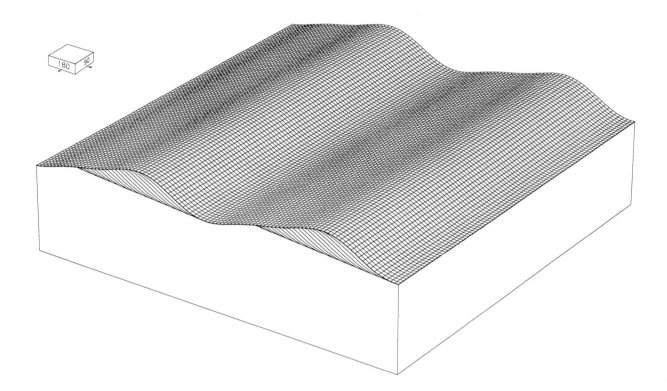

FIG. 9.— Structure formed by bedforms climbing at a stoss-erosional, lee-depositional, net-negative angle of climb.

RECOGNITION: When bedforms climb at a negative angle, they scour a wavy surface into the underlying substrate. The spacing of the undulations on the bounding surface is equal to the bedform spacing, but the complete bedform height is not necessarily preserved. Bedforms climbing at negative angles have cores that are composed in part of older substrate. As shown in Figure 5, bounding surfaces scoured by positively climbing bedforms dip upcurrent relative to the depositional surface. In contrast, bounding surfaces scoured by bedforms climbing at an angle of $0°$ are parallel to the depositional surface, and bounding surfaces scoured by negatively climbing bedforms dip downcurrent (relative to the depositional surface).

ORIGIN: In areas undergoing net erosion, bedforms migrate downward relative to the generalized depositional surface. This downward scouring—climbing at a negative angle—is probably just as common as climbing at a positive angle, but, because the process is erosional rather than depositional, the resulting structures are under-represented in the geologic record. This kind of structure can originate at all scales and in any environment where two-dimensional bedforms exist. Small-scale examples formed by ripples can be preserved where the bedforms are later buried by deposition. Large-scale examples can be shown to occur at the surface of some bedform fields such as in the Strzelecki and Simpson deserts in Australia, where dunes are composed in part of non-eolian substrate (Folk, 1971; Breed and Breed, 1979).

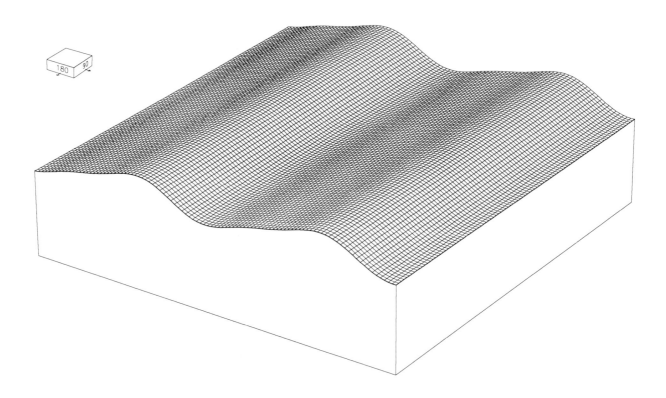

FIG. 10.— Structure formed by a stoss-erosional and lee-erosional angle of climb.

RECOGNITION: This structure is purely erosional, and consequently is probably extremely rare. Preservation can occur when the rate of deposition increases, a similar situation to that shown in Figures 13 and 14. Bedform spacing is indicated by the spacing of undulations on the bounding surface.

ORIGIN: Where the rate of erosion approaches the rate of bedform migration, bedforms scour downward into the underlying substrate without accumulating sediment, even on their lee sides. As with the high positive angles of climb, formation of this structure requires exceptional circumstances: a flow that can erode sediment without causing bedform migration. In general, these conditions are favored by longitudinal bedforms and by flows that accelerate downcurrent. The downcurrent acceleration means that the flow is able to transport more sediment than it contains, and a longitudinal bedform trend tends to prohibit local deposition by eliminating or restricting sites where the flow decelerates over the bedform surface. Examples of this kind of structure occur in air and water. Yardangs are possible eolian examples, and sediment furrows that occur in the deep sea and in estuaries are possible subaqueous examples. Oscillation ripples that occur in an area undergoing erosion can also produce this structure (Fig. 11).

FIG. 11.— Structure formed by ripples climbing at a lee-erosional angle; fluvial deposits of the Colorado River, Grand Canyon National Park, Arizona.

RECOGNITION: The rippled surface at the top of the light-colored sand was formed when ripple migration was accompanied by net erosion; the ripples scoured into the underlying flat-bedded sand. This erosional structure was preserved when the rate of deposition became positive, thereby causing the ripples to climb at a positive angle. The change from net erosion to net deposition coincided with an increase in silt content in the sediment. In the Grand Canyon, silt content increases during floods, suggesting that the change from erosion to deposition at this site was caused by a pulse of silty flood sediment.

Variable Two-Dimensional Bedforms and Cross-Bedding

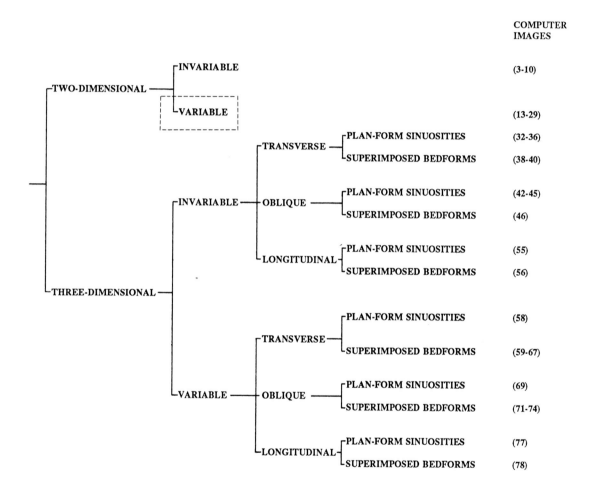

FIG. 12.— Schematic diagram showing the sequence in which illustrations are presented. Dashed line shows which structures are included in the following section.

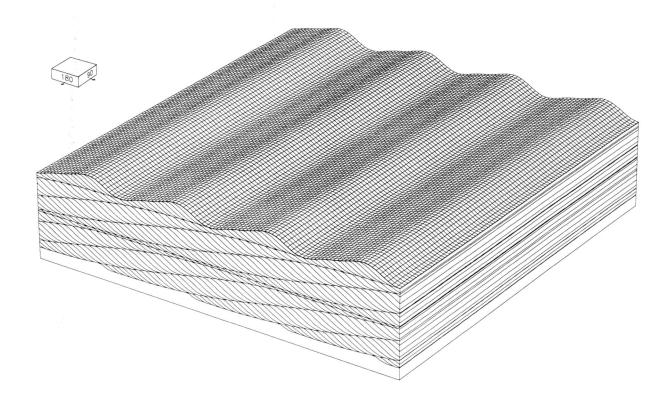

FIG. 13.— Structure formed by bedforms climbing at an angle that is fluctuating but always positive. This structure is an example of Allen's (1982) pattern IV ripple climb.

RECOGNITION: This structure is very useful for reconstructing bedform morphology, because the bedding indicates the plane of the generalized depositional surface. Although the complete depositional surface is not preserved unless the bedforms climb at a stoss-depositional angle, the generalized depositional surface is approximated by the plane at which the angle of climb simultaneously changed for adjacent bedforms (Rubin and Hunter, 1982). The bedform spacing is indicated by the distance between bounding surfaces, measured in a plane parallel to the generalized depositional surface and in a direction normal to the lines of intersection of cross-beds and bounding surfaces. As in other structures formed by variable two-dimensional bedforms, the cross-bed and bounding-surface dips plot along a single straight line through the center of the plot.

ORIGIN: This structure forms as the result of depositional episodes that cause the rate of deposition to increase and then decrease relative to the rate of bedform migration. Suitable depositional events can originate in rivers (Fig. 14), eskers (Allen, 1972), density currents, and eolian flows (Rubin and Hunter, 1982, fig. 4D). The events can be caused by increases and decreases in flow velocity or by increases and decreases in sediment availability upstream from the depositional site.

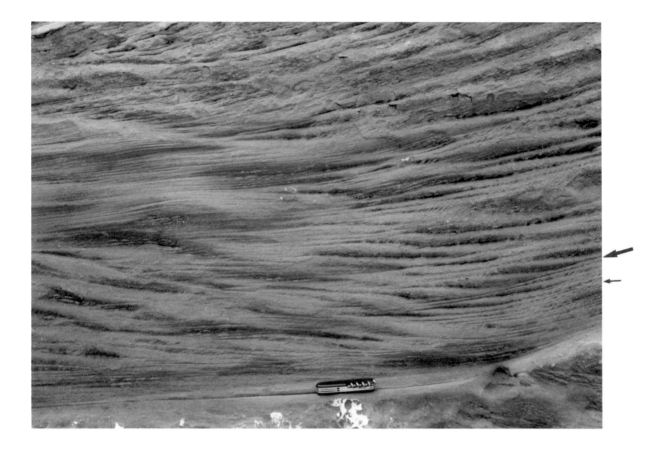

FIG. 14.— Structures produced by ripples with a fluctuating positive angle of climb; fluvial beds from the Kayenta Formation (Upper Triassic?) in Zion National Park, Utah.

RECOGNITION: This photograph shows approximately a dozen cycles of fluctuating angle of climb. Ripple foresets are visible at many locations on the right side of the photograph (small arrow) and indicate that the direction of ripple migration was from right to left. As illustrated by the computer-generated version of this structure (Fig. 13), increases in the rate of deposition relative to the rate of bedform migration cause the ripples to climb more steeply and to deposit thicker cross-laminated beds; decreases in the angle of climb produce thinner cross-laminated beds with bounding surfaces that more nearly parallel the generalized depositional surface. The generalized depositional surface (indicated by planes along which adjacent ripples simultaneously changed their angle of climb) dipped toward the left and steepened during deposition of the cyclic beds. One of the many depositional surfaces recognizable by a change in angle of climb is indicated by the large arrow at the right.

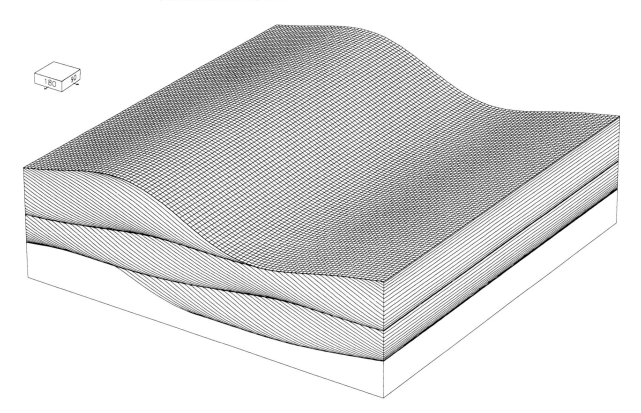

FIG. 15.— Structure formed by bedforms undergoing small and slow fluctuations in height. The left figure shows the bedform surface at a time when bedform height was a maximum; the right figure shows the same bedform at a later time when height was a minimum in the height-fluctuation cycle.

RECOGNITION: Unless the rate of deposition is great enough to provide all the sediment necessary to make the bedforms larger, an increase in height must be accompanied by transfer of sediment from bedform troughs to their crests. The rate of deposition usually is not great enough, and fluctuations in bedform height cause the elevation of the bedform troughs to rise and fall (Terwindt, 1981). The resulting structures have gently undulating lower bounding surfaces, as shown here, or have scalloped bounding surfaces, as shown in Figures 16 and 17.

ORIGIN: Fluctuations in bedform height can be random changes undergone by individual bedforms or, as illustrated here, can be systematic changes undergone by entire populations of bedforms. Systematic fluctuations in height of such bedforms as sand waves can arise from changes in flow velocity or flow depth (Rubin and McCulloch, 1980) and can probably arise from fluctuations in flow direction. Cyclic fluctuations in height of sand waves have been observed to result from neap-spring fluctuations in velocities of tidal currents (Boersma and Terwindt, 1981; Dalrymple, 1984; Terwindt and Brouwer, 1986), and fluctuations in sand-wave height in rivers have been observed to result from fluctuations in discharge (Coleman, 1969). In contrast to the systematic cyclic fluctuations in bedform height, which require cyclic fluctuations in flow, random fluctuations in height undergone by individual bedforms are probably extremely common in all environments, even in steady flows.

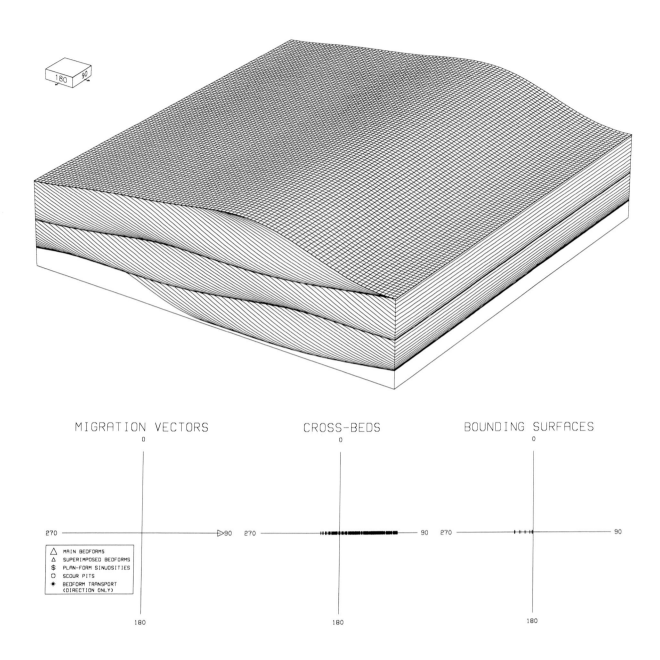

MIGRATION VECTORS

CROSS-BEDS

BOUNDING SURFACES

△ MAIN BEDFORMS
△ SUPERIMPOSED BEDFORMS
$ PLAN-FORM SINUOSITIES
○ SCOUR PITS
✳ BEDFORM TRANSPORT
  (DIRECTION ONLY)

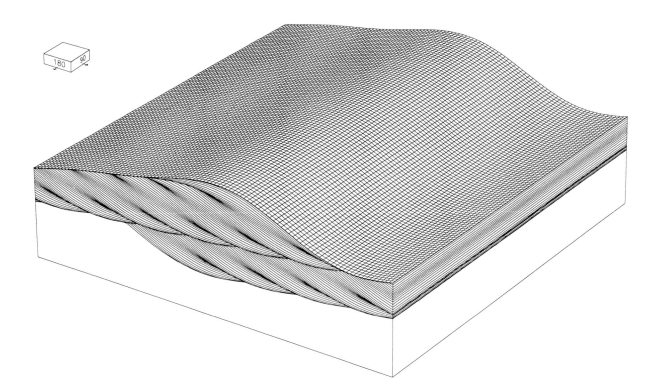

FIG. 16.— Structure formed by bedforms undergoing small but rapid fluctuations in height. This structure is one kind of scalloped cross-bedding (Rubin and Hunter, 1983; Rubin, 1987). The left figure shows the bedform surface at a time when bedform height was a maximum; the right figure shows the same bedform at a later time when height was a minimum in the height-fluctuation cycle.

RECOGNITION: Scalloped cross-bedding—compound cross-bedding with bounding surfaces that cyclically scoop down into previously deposited foresets or into sediment below the set—forms by cyclic fluctuations in bedform height (Figs. 16 and 17), by cyclic reversals in bedform asymmetry and migration direction (Figs. 21, 22, and 24), and by migration of superimposed bedforms over the lee slopes of larger bedforms (Figs. 46 and 71-74). The structures formed by bedforms fluctuating in height closely resemble—and may be indistinguishable from—structures formed by reversing bedforms. Like other structures formed by variable two-dimensional bedforms, the structure illustrated in this example has cross-bed and bounding-surface dips that plot along a single straight line.

ORIGIN: Although similar to the preceding example, this structure requires height-fluctuation cycles that are shorter relative to the bedform period (the time required for the bedforms to migrate a distance equal to the bedform spacing).

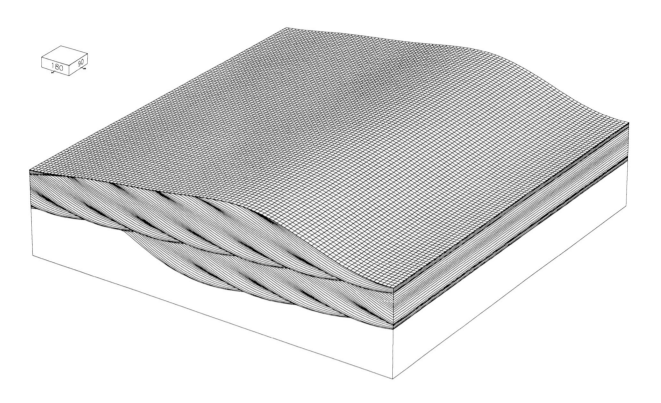

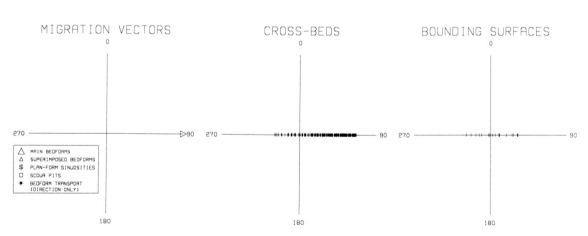

MIGRATION VECTORS

CROSS-BEDS

BOUNDING SURFACES

△ MAIN BEDFORMS
△ SUPERIMPOSED BEDFORMS
$ PLAN-FORM SINUOSITIES
○ SCOUR PITS
✳ BEDFORM TRANSPORT
(DIRECTION ONLY)

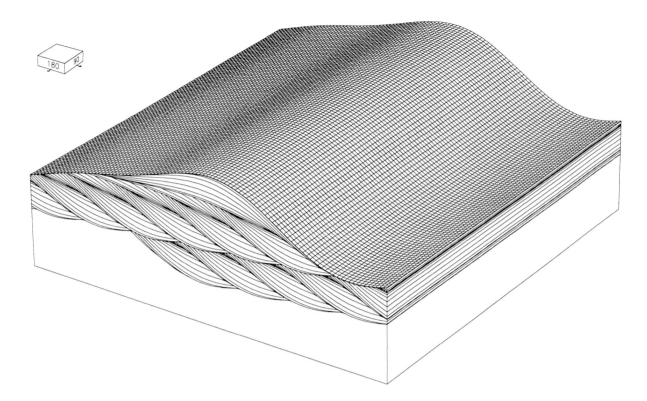

FIG. 17.— Structure formed by bedforms undergoing large and rapid fluctuations in height. The left figure shows the bedform surface at a time when bedform height was a maximum; the right figure shows the same bedform at a later time when height was a minimum in the height-fluctuation cycle.

RECOGNITION: This structure is similar to the preceding example, but the height fluctuations shown here are greater and have a more pronounced effect.

ORIGIN: This structure originates by the same process as the two preceding examples, but the greater height fluctuations in this example imply a greater fluctuation in flow conditions.

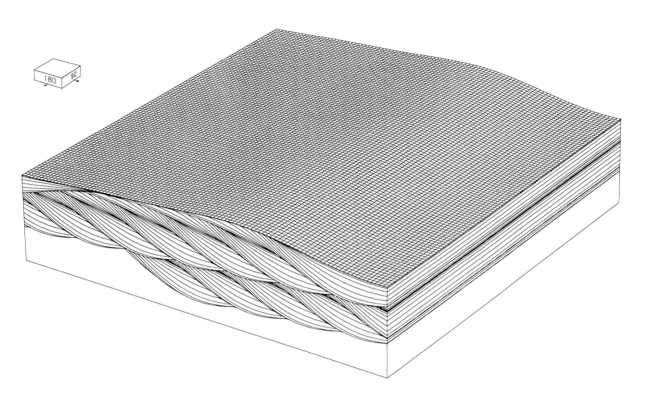

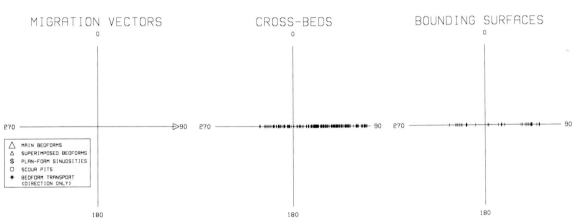

MIGRATION VECTORS

CROSS-BEDS

BOUNDING SURFACES

△ MAIN BEDFORMS
△ SUPERIMPOSED BEDFORMS
$ PLAN-FORM SINUOSITIES
○ SCOUR PITS
✳ BEDFORM TRANSPORT
  (DIRECTION ONLY)

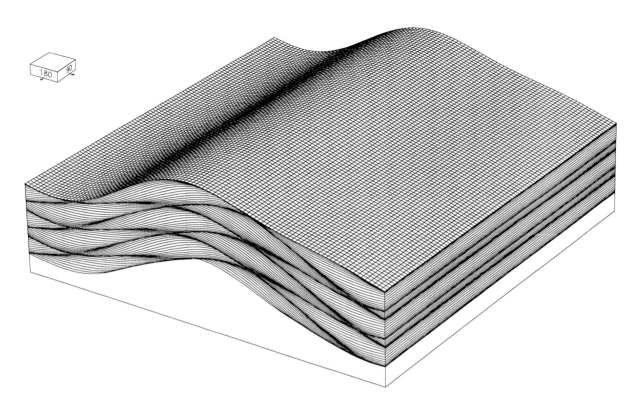

FIG. 18.— Structure formed by bedforms that reverse asymmetry without net displacement. The left illustration shows the bedform with its most extreme left-facing asymmetry, and the right illustration shows the bedform at a later time with its most extreme right-facing asymmetry.

RECOGNITION: In many natural flows, reversals in asymmetry commonly occur simultaneously with reversals in migration direction because both kinds of reversals are caused by reversals in flow direction. These two responses are simulated separately, because, for reasons discussed below, the two kinds of bedform reversals can occur independently in some flows. The structure shown here is similar to those produced by bedforms that reverse migration direction (Fig. 19); both have erosion surfaces that are arranged in vertically zig-zagging sequences. The two kinds of structures can be distinguished, however, because reversals in asymmetry (shown here) cause cross-beds to offlap and onlap the erosion surfaces, whereas reversals in migration direction (Fig. 19) produce erosion surfaces with relatively concordant overlying beds. Reversals in asymmetry also produce zig-zags in the centers of the bedform troughs—a feature that is not produced by reversals in migration direction.

Vertical profiles of cross-bed dips vary with location in this structure. Profiles through crest or trough deposits have cross-bed dips that reverse direction, whereas dips through flank deposits vary only in inclination.

ORIGIN: The response of bedforms to reversing flows is a complicated problem because flow reversals can occur on more than one time scale in a single flow, and the different scales of flow reversals have different effects on the bedforms. Some flow reversals can cause reversals in asymmetry without net migration, and other flow reversals can cause reversals in migration direction without causing significant asymmetry. For example, in oscillatory flows where the volume of sediment transported during each oscillation approaches the volume of sediment in the bedforms, the bedforms will reverse asymmetry. Where such an oscillatory flow is symmetrical, the bedforms will not undergo net migration. In contrast, where the volume of sediment transported during individual oscillations is small relative to the volume in the bedforms, the bedforms maintain a symmetrical profile. Inducing a slight asymmetry to the oscillatory flow will cause the bedforms to migrate while maintaining a relatively symmetrical profile, and reversing the direction of net asymmetry in the flow will cause the nearly symmetrical bedforms to reverse migration direction. Distinguishing the structures formed by the different kinds of bedform reversals (asymmetry or migration direction) is, therefore, important because the structures indicate flow reversals on different time scales. Real examples of structures indicating flow

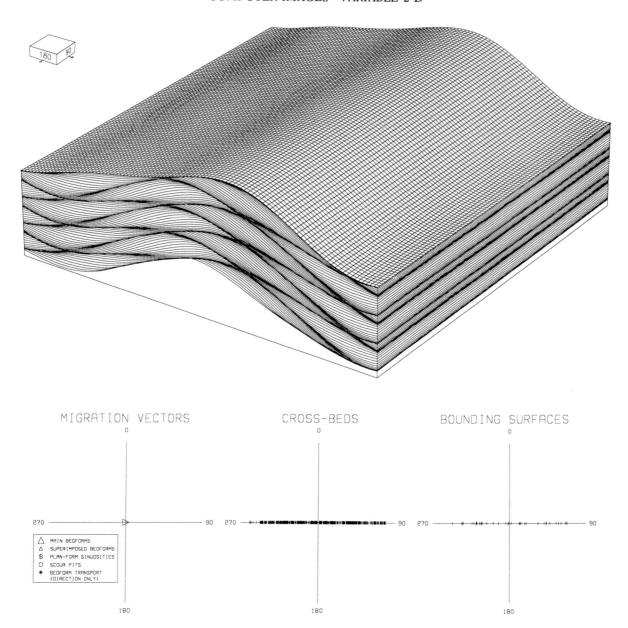

reversals on more than one time scale are shown in Figures 20 and 64.

The structure shown in this computer image was produced by vertically climbing bedforms. The structure is relatively rare for the same reasons that other vertically climbing bedforms are rare (Fig. 3). Even in oscillatory flows that might be expected to produce structures like the one shown here, sand transport in opposing directions tends to be imbalanced, thereby causing bedforms to migrate (Newton, 1968). Bagnold (1941) predicted that longitudinal dunes would have the kind of internal structure illustrated in this example, but eolian deposits having this structure have been rarely—if ever—documented in the geologic record (Rubin and Hunter, 1985). This kind of structure, however, could be expected to form in convergence zones in reversing flows such as might occur at isolated locations on some tidal ridges, at eddy-reattachment zones along river banks, at restricted locations in fields of linear eolian dunes or symmetrical sand waves, or at restricted sites in wave-generated flows.

39

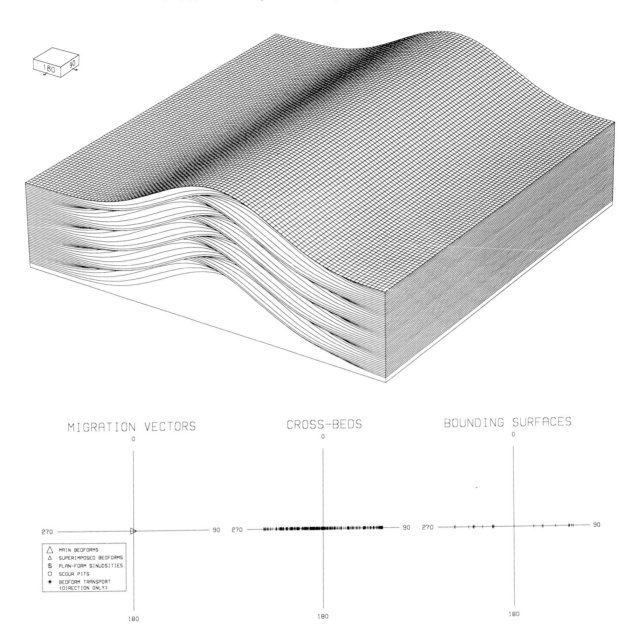

FIG. 19.— Structure produced by symmetrical bedforms that migrate back and forth a fraction of the distance of the bedform spacing but have no net displacement.

RECOGNITION: This structure is similar to those produced by bedforms that reverse asymmetry (Fig. 18). The two structures can be distinguished, however, because reversals in asymmetry cause cross-beds to offlap and onlap the erosion surfaces and cause zig-zag structures within the troughs. In contrast, the reversals in migration direction shown here produce erosion surfaces with relatively concordant overlying beds. The two kinds of structures may be distinguishable in good outcrops, but the cross-bed-dip patterns of the two structures are virtually indistinguishable.

ORIGIN: Deposition of this structure requires a flow that maintains a constant—probably symmetrical—bedform shape, while simultaneously causing the bedforms to reverse their direction of migration. These conditions are most likely to be met in oscillatory flows in which the amount of sediment transported during individual flow cycles is small relative to the amount contained in the bedforms. At the same time, the conditions for vertical climb (Fig. 3) must be satisfied. A fluvial example is shown in Figure 20.

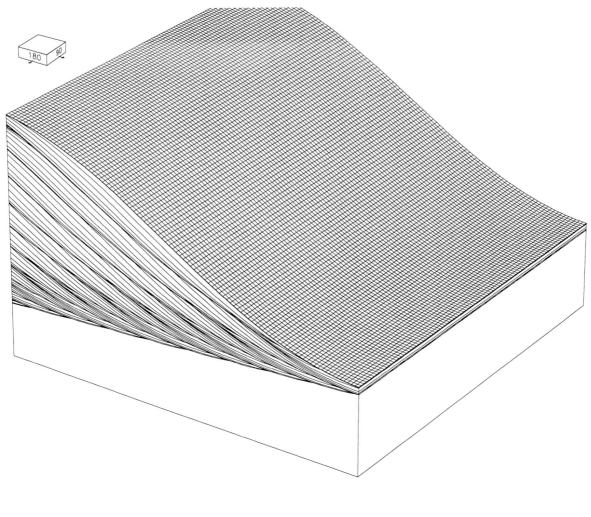

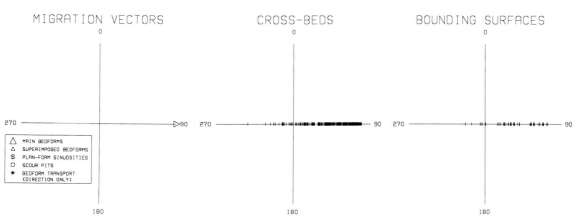

MIGRATION VECTORS   CROSS-BEDS   BOUNDING SURFACES

△ MAIN BEDFORMS
△ SUPERIMPOSED BEDFORMS
$ PLAN-FORM SINUOSITIES
○ SCOUR PITS
✳ BEDFORM TRANSPORT
  (DIRECTION ONLY)

B

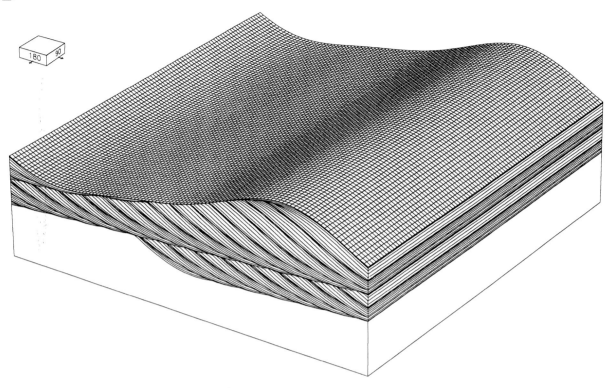

FIG. 22.—*Continued*

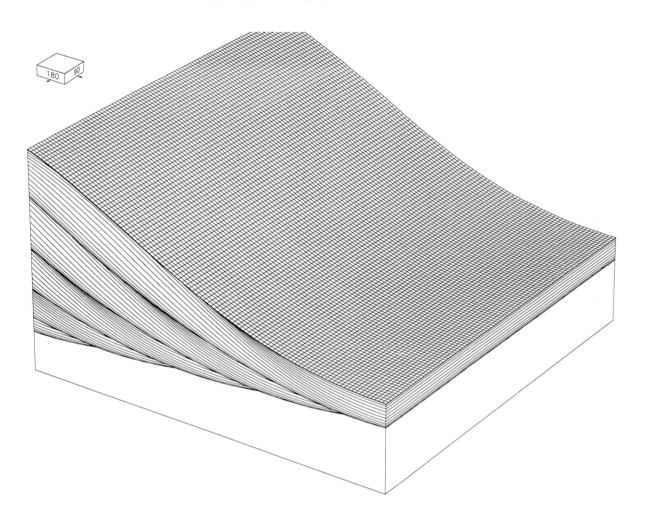

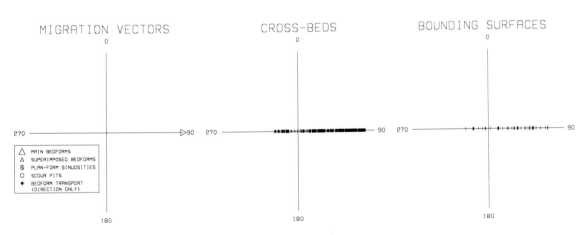

MIGRATION VECTORS

CROSS-BEDS

BOUNDING SURFACES

△	MAIN BEDFORMS
△	SUPERIMPOSED BEDFORMS
$	PLAN-FORM SINUOSITIES
O	SCOUR PITS
✳	BEDFORM TRANSPORT (DIRECTION ONLY)

FIG. 23.— Structure inferred to have formed by a dune that fluctuated in asymmetry and migration speed; eolian deposits in the Cedar Mesa Sandstone Member (Permian) of the Cutler Formation, southeast Utah.

RECOGNITION: Cyclic foresets, such as those in this example, clearly indicate cyclic depositional processes. The bedding in this set of cyclic foresets has a characteristic that suggests that the cyclicity was caused by fluctuating flow: wedges of sediment deposited along the lee slope (light-colored bottomset and foreset beds). Deposition of these basal wedges suggests that the cyclicity was produced by fluctuations in asymmetry and migration speed, as simulated in Figure 22A.

FIG. 24.— Scalloped cross-bedding inferred to have been produced by a dune undergoing cyclic fluctuations in height or asymmetry and migration speed; eolian deposits in the Navajo Sandstone (Upper Triassic? and Jurassic) in Water Holes Canyon, Arizona.
RECOGNITION: Measurements in the field show that the cross-beds and bounding surfaces in this cross-stratified bed have the same strike, thereby demonstrating that the cyclicity of the bedding was produced by cyclic flows rather than superimposed bedforms. The cyclic flows are inferred to have caused the dunes to vary in height, as illustrated in Figure 16, or to vary in asymmetry and migration speed as is illustrated in Figure 22B.

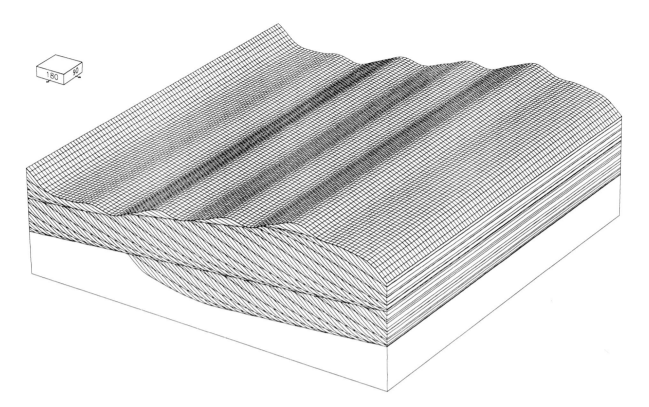

FIG. 25.— Structure formed by bedforms with parallel superimposed bedforms migrating in the same direction.

RECOGNITION: This structure closely resembles other kinds of cross-bedding with cyclic foresets, but is perhaps distinguishable because the cross-beds deposited by the superimposed bedforms consistently downlap along the bounding surfaces scoured by the superimposed bedforms. In contrast, cross-bedding formed by reversals in migration direction or fluctuations in bedform morphology typically contains basal wedges (Fig. 22A) or scalloped cross-bedding with relatively conformable upcurrent-dipping beds that immediately overlie the lower set boundary (Fig. 22B). In more realistic depositional situations, either real or computer-generated, superimposed bedforms are unlikely to be exactly parallel to the main bedforms for long distances along-crest, and recognizing the deposits of downslope-migrating bedforms is easier than in the simulation shown here. Local differences in orientation of the crests of the two sets of bedforms cause the cross-beds that are deposited by the superimposed bedforms to dip in a different direction from the bounding surfaces that are scoured by the superimposed bedforms (Figs. 65 and 66). The polar plot of this structure (not shown) is similar to the plots of all other two-dimensional cross-bedding: cross-bed and bounding-surface poles plot along a straight line through the center of the plot.

ORIGIN: All of the previous examples of structures that form by cyclic fluctuations in bedform morphology or path of climb require flows that cyclically fluctuate in strength or direction. In the structure illustrated in this example, however, bedform morphology fluctuates even in steady flows. The fluctuations in morphology result from the differing migration speeds of the two sets of bedforms. Cyclic passage of the superimposed bedforms over the main bedforms causes cyclic constructive and destructive interference, thereby generating cyclic foresets even in steady flows.

Superpositioning of bedforms is a common phenomenon in most flows where bedforms are large enough for other bedforms to be accommodated on them. The structure illustrated here is an approximation of structures that are found in fluvial deposits (Banks, 1973; McCabe and Jones, 1977), tidal deposits (Dalrymple, 1984), and eolian deposits (Brookfield, 1977; Fig. 26), but in most natural examples the superimposed bedforms could be expected to have crestlines with a slightly different trend, sinuosity, or along-crest length relative to the main bedform, as illustrated in many of the following computer images.

FIG. 26.— Cyclic compound cross-bedding inferred to have been produced by small dunes migrating down the lee slope of a larger dune; Navajo Sandstone (Upper Triassic? and Jurassic), Zion National Park, Utah.

RECOGNITION: Lack of basal wedges, lack of upcurrent-dipping cross-beds near the base of the set, and lack of upslope-migrating superimposed ripples suggest that this deposit was formed by downslope-migrating superimposed bedforms (Fig. 25) rather than by cyclically reversing flows (Fig. 22). If this interpretation is correct, field measurements should show that the cross-beds deposited by the superimposed bedforms dip in slightly different directions from the bounding surfaces scoured by the superimposed bedforms, as illustrated in Figures 65 and 66; cross-beds and bounding surfaces produced by superimposed bedforms can have the same dip direction only if crestlines of the superimposed bedforms exactly parallel the crestline of the main bedform (Fig. 25).

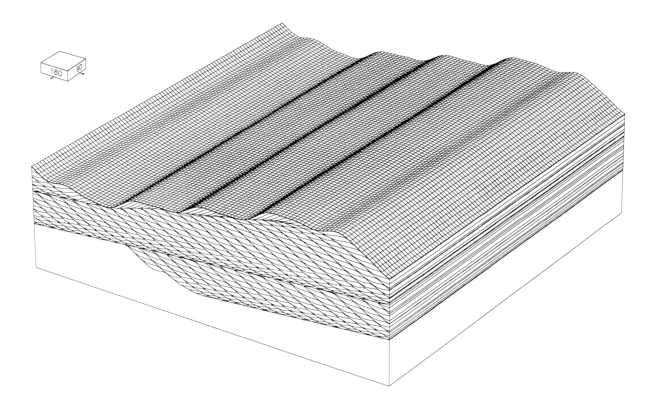

FIG. 27.— Structure formed by bedforms with superimposed bedforms migrating in the opposite direction. **RECOGNITION:** This structure differs from the preceding example in having cross-beds that consistently onlap the bounding surfaces scoured by the superimposed bedforms.

ORIGIN: As in the preceding example, this structure forms without requiring fluctuating flows. Structures are unlikely to form exactly as illustrated in this example, because the two sets of bedforms are migrating in opposite directions across the entire bed surface at all times. Superimposed bedforms can migrate up the lee slope of the main bedform in at least two situations, however. First, upcurrent migration of superimposed bedforms occurs locally in the troughs of bedforms where the lee eddy drives superimposed bedforms in an upcurrent direction (Boersma and others, 1968; Dalrymple, 1984). Second, upcurrent migration of superimposed bedforms occurs temporarily on some bedforms when flow reverses, as in tidal flows (Terwindt, 1981; Figs. 29 and 30) and within river eddies (Fig. 28). The depositional situation shown here includes upcurrent-migrating stoss-side superimposed bedforms, because the computer program is not capable of causing the migration direction of the superimposed bedforms to vary with location on the main bedform. This does not affect the geometry of the structure, however, because the stoss-side deposits are not preserved.

FIG. 28.— Structure produced by a migrating bedform with superimposed bedforms that migrated down and up its lee slope; fluvial deposits, Colorado River, Grand Canyon National Park, Arizona. Area shown is 35 cm from top to bottom. This structure is a combination of the structures simulated in Figures 25 and 27 and also includes structures produced by bedforms climbing at stoss-depositional angles.

RECOGNITION: While the main bedform that deposited these beds migrated from right to left, ripples that were superimposed on its lee slope repeatedly reversed their direction of migration (A). Eventually the ripples were replaced with a larger upslope-migrating bedform that deposited a relatively thick set of foresets (B). The main bedform continued to migrate to the left after the superimposed bedforms disappeared (C). If found in the geologic record, these beds might be incorrectly identified as tidal deposits (Figs. 29 and 30), because of the flow reversals indicated by the reversals in ripple-migration direction. The real cause of the flow reversals was probably the formation and decay of eddies at the depositional site. The eddies may have been restricted to the lee side of the main bedform or may have been more widespread eddies in the lee of river-channel constrictions (Schmidt, 1986).

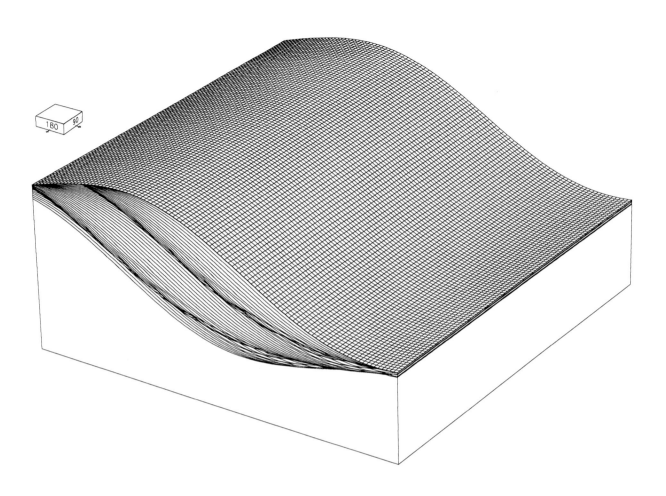

FIG. 29.— Simulated tidal bundles. Input parameters were adjusted so that the bedforms would behave like the tidal bedforms described by Visser (1980), Boersma and Terwindt (1981), and Terwindt (1981). With each diurnal tide cycle, the simulated sand waves vary in asymmetry and migration speed; superimposed ripples develop during times of weak currents. The ripples migrate faster than the main bedforms and reverse migration direction when the tidal flow reverses.

RECOGNITION: Fluctuating flow is indicated by the reversals in migration direction of the superimposed ripples. Cyclicity of the flow fluctuations is demonstrated by the cyclic spacing of bounding surfaces that define the bundles.

ORIGIN: This structure requires cyclic fluctuations in flow velocity and direction. The best examples of this kind of structure occur in tidal deposits (Fig. 30), but cyclic foresets produced by annual wind cycles occur in eolian deposits (Figs. 23 and 24).

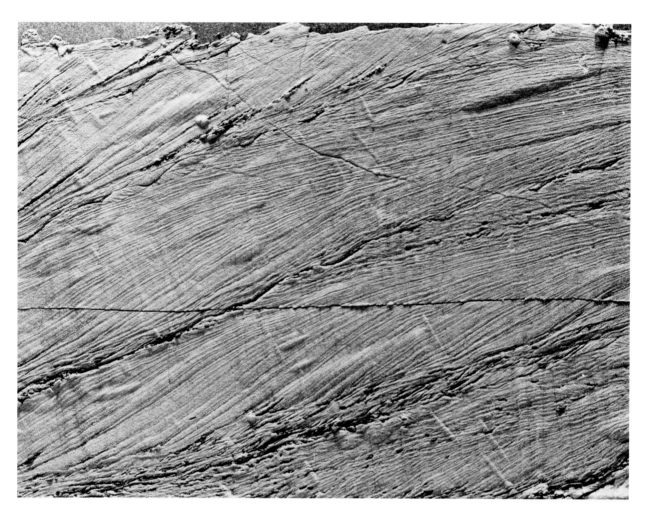

FIG. 30.— Tidal bundles; modern deposits from the Oosterschelde, The Netherlands, photographed by Joost Terwindt. Set is approximately 1 m thick.

RECOGNITION: As in the computer-generated version of this kind of structure (Fig. 29), ripples migrate up the lee slope, slack water occurs, and then ripples reverse migration direction. In addition to this geometric evidence of flow reversals, cyclic fluctuations in flow velocity are indicated by the cyclically spaced mud drapes. Additional details of the origin and interpretation of this structure are given by Terwindt (1981, fig. 4).

Invariable Three-Dimensional Bedforms and Cross-Bedding

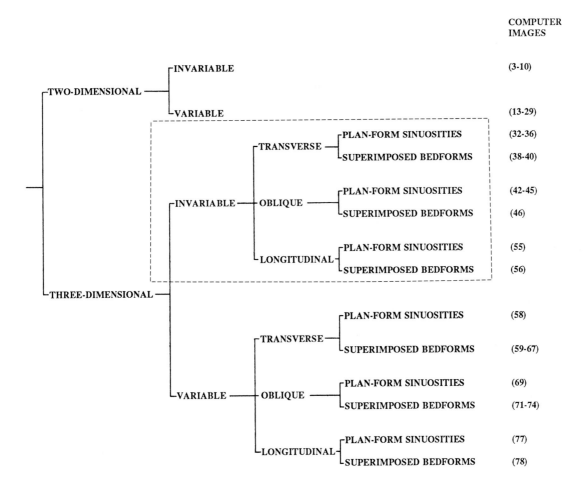

FIG. 31.— Schematic diagram showing the sequence in which illustrations are presented. Dashed line shows which structures are included in the following section.

A

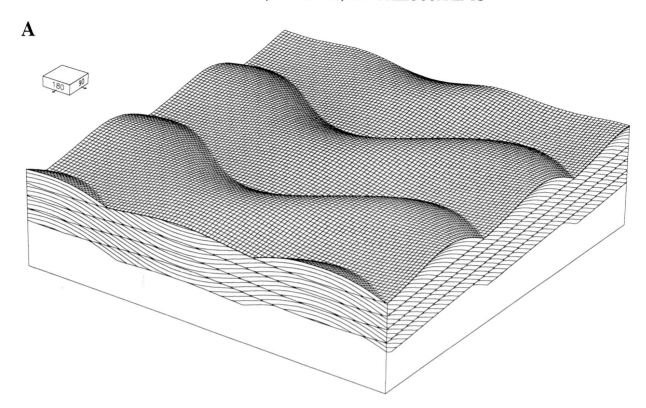

FIG. 32.— Structures formed by transverse bedforms with curved, in-phase crestlines: (A) sine-shaped plan form, (B) linguoid plan form, and (C) lunate plan form.

RECOGNITION: The sinuous plan-forms of these bedforms cause the foresets that are produced to dip toward a wide range of directions. The cross-bed-dip plots consist of radial lines, because the direction of dip is constant throughout each vertical profile. The in-phase character of the crestlines, and the resulting lack of scour pits in the bedform troughs, cause the bounding surfaces that are scoured to be nearly planar. The slight curvature of the bounding surfaces results from the differing locations (in a downcurrent direction) at which the deepest point in the trough occurs. As the angle of climb approaches 0°, bounding surfaces become more nearly planar. This structure is easily distinguished from bedforms with out-of-phase crestlines, because those bedforms have scour pits in their troughs and, as a result, produce trough-shaped sets of cross-beds (Fig. 34).

The transverse orientation of these bedforms relative to the flow direction is indicated in horizontal sections by an absence of along-crest displacement of crestline sinuosities. In structures where such along-crest displacement is present, it is visible in sections parallel to the depositional surface (horizontal sections in Figs. 42-43). Such displacement also causes an asymmetrical distribution of dip directions of cross-beds and bounding surfaces that is visible in crest-parallel vertical sections and in plots of dip directions. Structures formed by transverse in-phase bedforms with sine-shaped, linguoid, and lunate crest plans are so similar that, except in unusually extensive horizontal sections, distinguishing their deposits is likely to be impossible.

The wiggles in some of the cross-bed traces on the horizontal surface are not real. They are an artifact of the contouring program, and they are produced where the foresets are nearly horizontal.

ORIGIN: Bedforms with curved in-phase crestlines are relatively two-dimensional and could be expected to form in flows similar to those that produce two-dimensional bedforms. The specific hydraulic requirements for curved in-phase crestlines are not yet known.

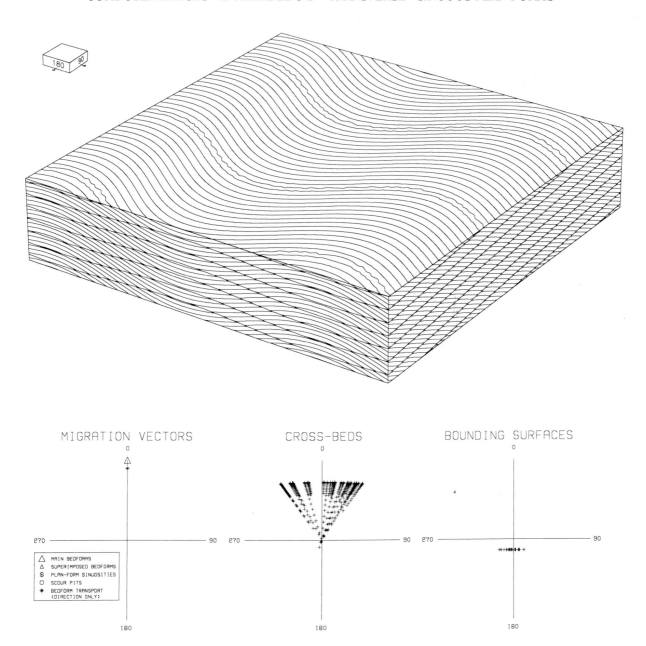

MIGRATION VECTORS

CROSS-BEDS

BOUNDING SURFACES

△ MAIN BEDFORMS
△ SUPERIMPOSED BEDFORMS
$ PLAN-FORM SINUOSITIES
○ SCOUR PITS
✳ BEDFORM TRANSPORT
  (DIRECTION ONLY)

B

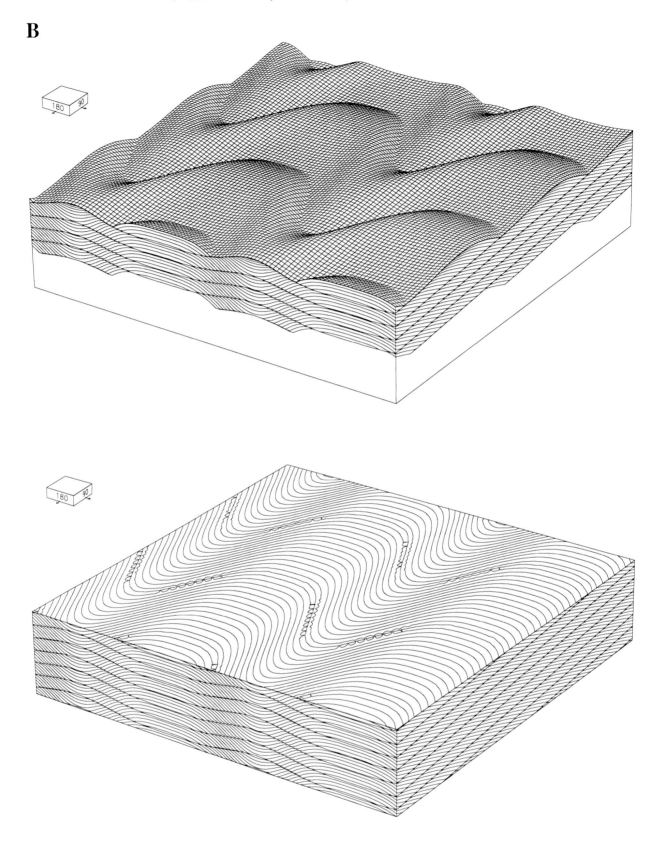

FIG. 32.—*Continued*

FIG. 37.— Relatively complicated cross-bedding formed by irregular, three-dimensional dunes; eolian deposits in the Temple Cap Sandstone (Jurassic), Zion National Park, Utah. This photograph shows a real example of the kind of structure simulated in Figure 36. The sets of cross-beds are as much as several meters thick.

RECOGNITION: The lack of a regular pattern to this cross-bedding suggests that the cross-stratified beds were deposited by bedforms with different shapes or different phase relations between crestlines. Trough-shaped sets form where adjacent bedform crestlines are locally out of phase (Fig. 34), whereas more tabular sets form where adjacent bedforms are in phase (Fig. 32). Changes through time may have occurred during deposition of these beds but cannot be demonstrated from the photograph.

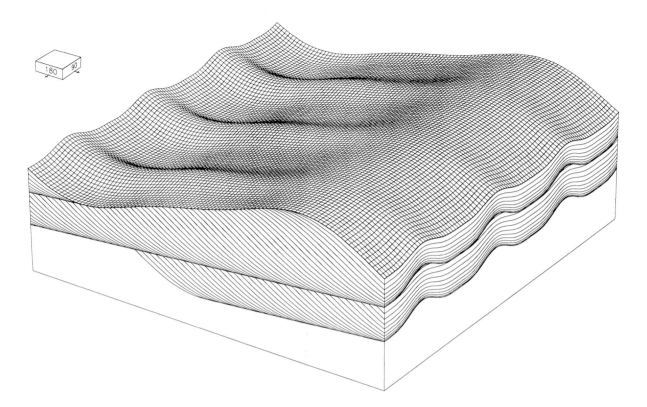

FIG. 38.— Structure formed by transverse, straight-crested bedforms with superimposed features that might be described either as longitudinal spurs or symmetrical longitudinal bedforms.

RECOGNITION: Transverse orientation of the bedforms that produced these structures is indicated in the horizontal section, which shows that scour pits and spurs did not migrate along the crests of the larger bedforms on which they were superimposed. The trend of the main bedforms is represented by the generalized strike of the cross-beds and by an imaginary line connecting the "fingertips" formed by migration of adjacent scour pits upward through the horizontal outcrop plane. The direction of scour-pit migration, which in this example is normal to the bedform crestline, is indicated by the axes of the troughs that were scoured by the migrating scour pits. The longitudinal trend of the superimposed spurs can also be demonstrated by their vertical accretion (manifested as symmetrical convex-up structures in vertical section). Transverse orientation of the main bedforms and lack of along-crest migration of the spurs result in symmetrical filling of scour pits and symmetrical distribution of cross-bed poles relative to bounding-surface poles.

Because the three-dimensionality of these bedforms arises from superpositioning of the longitudinal spurs, bedform troughs vary in elevation from spur to scour pit, and cross-beds are in phase and conformable with bounding surfaces. Bedforms that are three-dimensional because of sinuous crestlines do not produce such structures (Figs. 32 and 42). The effect of crest-plan sinuosities is also distinguishable from that of superimposed spurs or bedforms in polar plots of cross-bed dips. The former kind of three-dimensionality produces a fan-shaped distribution (Fig. 32), whereas the latter (or both kinds of three-dimensionality) produces scatter over a hemisphere, as shown here.

ORIGIN: As with other three-dimensional bedforms, the hydrodynamic cause of specific geometries is not yet understood.

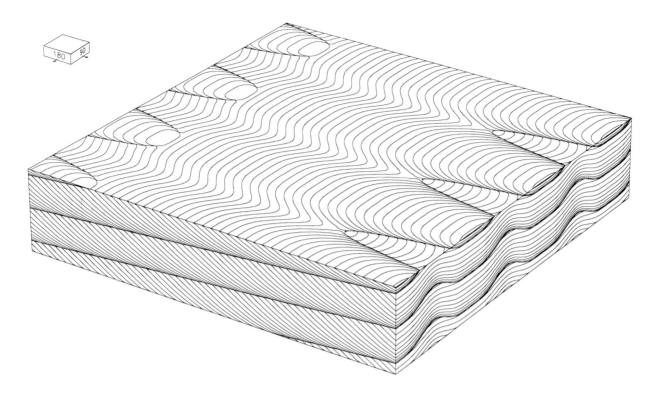

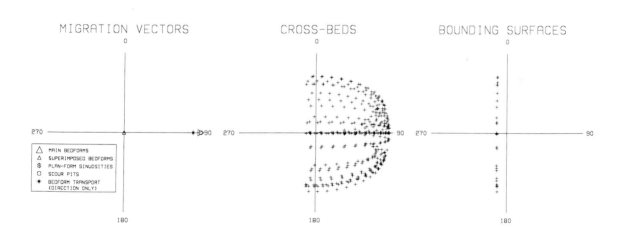

MIGRATION VECTORS

CROSS-BEDS

BOUNDING SURFACES

△ MAIN BEDFORMS
△ SUPERIMPOSED BEDFORMS
$ PLAN-FORM SINUOSITIES
O SCOUR PITS
✳ BEDFORM TRANSPORT
(DIRECTION ONLY)

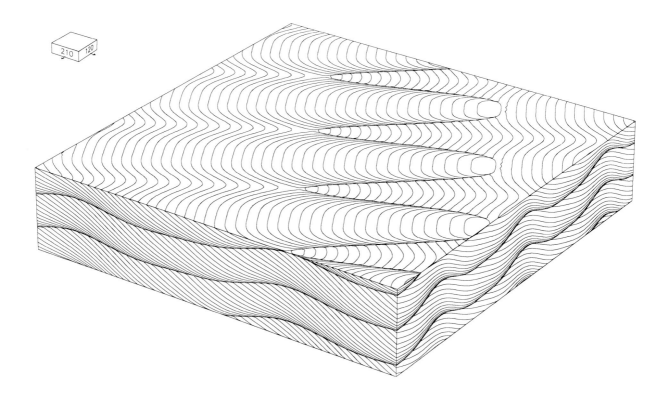

FIG. 38.—*Continued*

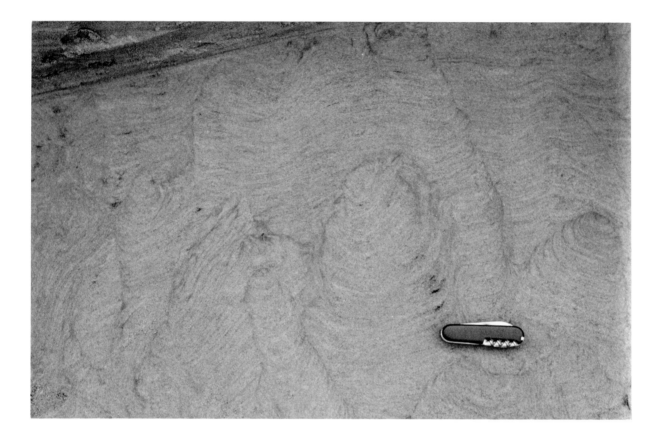

FIG. 39.— Horizontal section through the deposit of a bedform with superimposed longitudinal spurs and scour pits; fluvial deposits, Colorado River, Grand Canyon National Park, Arizona.

RECOGNITION: The trend of the ripples that deposited these rib-and-furrow structures was roughly from left to right, and the ripples migrated toward the top of the photograph. The two trough-shaped sets in the vicinity of the knife were filled relatively symmetrically and have axes that trend roughly normal to the strike of the somewhat straighter cross-beds at the top of the photograph, characteristics that are indicative of transverse bedforms (as simulated in Fig. 38). In contrast, two sets of cross-beds (to the left) have been preferentially truncated on their right sides; when systematic, such preferential truncation is more typical of oblique bedforms (Fig. 46).

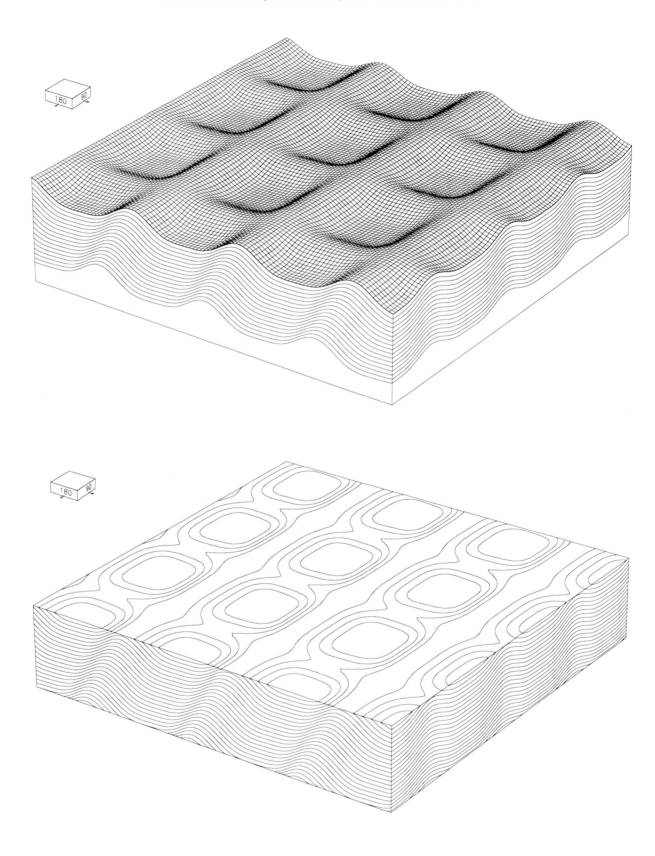

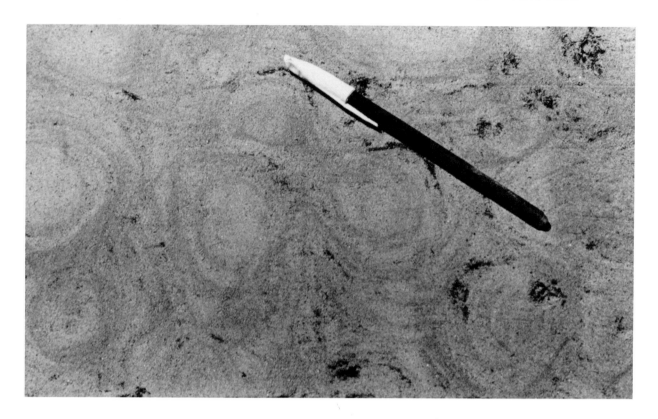

FIG. 40.—(At left) Structures formed by stoss-depositional, transverse bedforms with superimposed symmetrical bedforms or longitudinal spurs.

RECOGNITION: Despite containing scour pits in their troughs, these bedforms do not deposit trough-shaped sets of cross-beds, because the entire depositional surface is preserved. The superimposed spurs trend parallel to transport (and therefore do not migrate laterally). The scour-pit and spur deposits are stacked vertically as shown in the crest-parallel vertical section. Stoss-depositional climb of this bedform assemblage causes scour pits in the bedform troughs to migrate upward relative to the depositional surface, thereby producing distinctive structures characterized by cross-beds whose traces appear as nearly concentric circles in horizontal sections.

ORIGIN: As with other examples of stoss-depositional climb, formation of this structure requires relatively rapid rates of deposition. Otherwise, it is not as unusual a structure as might be imagined (Figs. 41 and 64). This kind of structure is discussed in greater detail in the exlpanations of Figures 63 and 64, reversing analogs of the structure shown here.

FIG. 41.— **(Above)** Horizontal section through structures formed by stoss-depositional ripples with spurs; fluvial deposits, Colorado River, Grand Canyon National Park, Arizona.

RECOGNITION: These structures are recognizable by their distinctive circular cross-bed traces. The trend of the ripples was rather irregular, as indicated by the variation in trend of the lines along which the circular cross-bed traces are arranged; one ripple crest is located under the pen. Displacement of the centers of some of the circular traces suggests that the ripples were asymmetric in profile or were migrating during deposition.

A

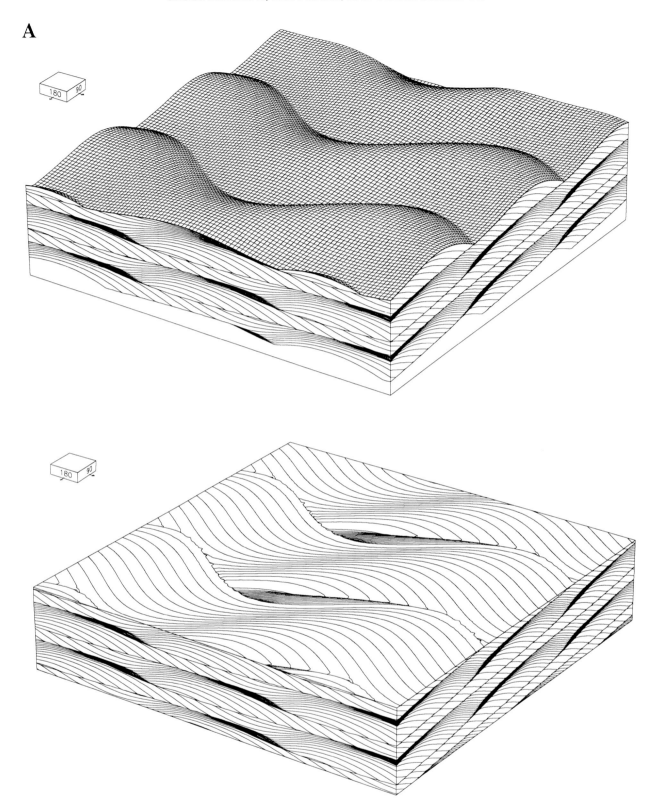

FIG. 42.— Structures formed by bedforms with along-crest-migrating in-phase sinuosities. (A) Rate of along-crest migration of sinuosities is equal to the rate of lateral migration of the main bedform. (B) Rate of along-crest migration of sinuosities is approximately four times the rate of lateral migration of the main bedform.

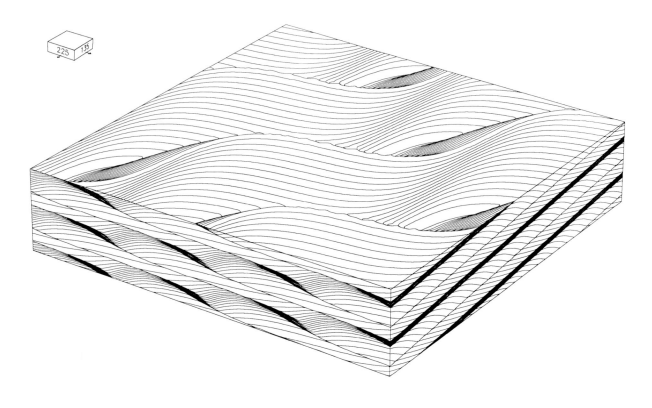

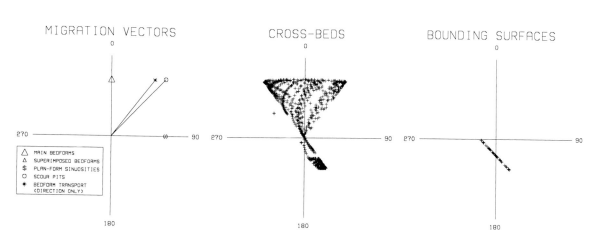

MIGRATION VECTORS

△ MAIN BEDFORMS
△ SUPERIMPOSED BEDFORMS
$ PLAN-FORM SINUOSITIES
O SCOUR PITS
* BEDFORM TRANSPORT
  (DIRECTION ONLY)

CROSS-BEDS

BOUNDING SURFACES

RECOGNITION: These structures are most easily recognized in horizontal sections, because such sections display both the lateral migration of the bedforms and the along-crest displacement of the crestline sinuosities. Like other structures deposited by oblique bedforms, this structure has dip patterns in which cross-beds and bounding surfaces are asymmetrically distributed relative to the center of the plot and relative to each other. Note the divergence between the trough axes and the cross-bed dip directions.

ORIGIN: Deposition of these structures requires sinuous bedforms that are oblique to the resultant transport direction. An example of real bedforms with along-crest-migrating sinuosities is shown in Figure 70.

B

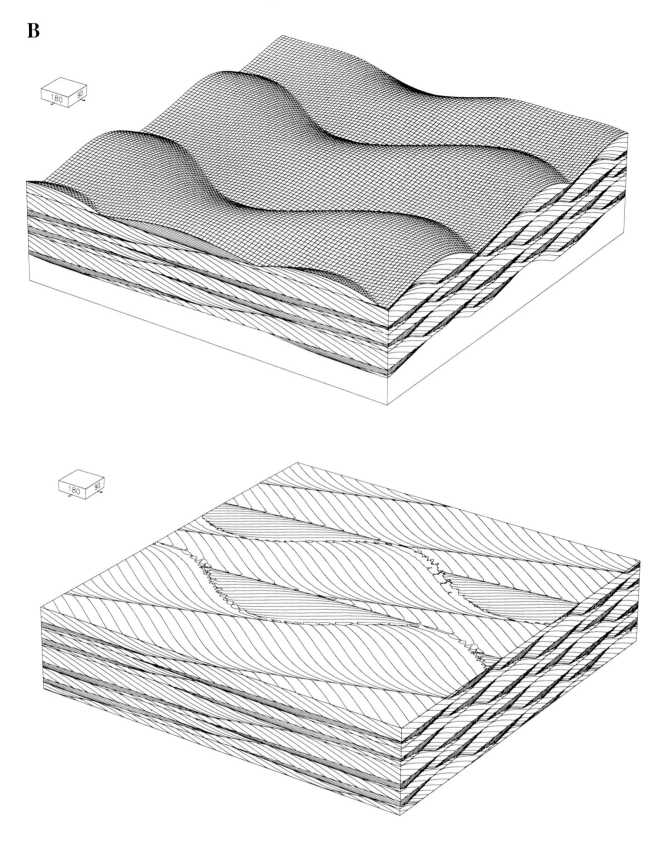

FIG. 42.—*Continued*

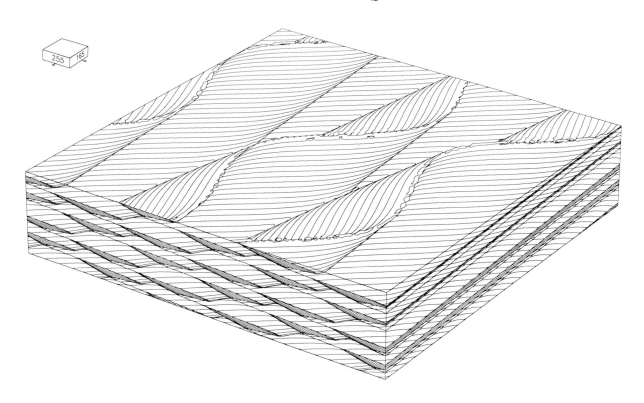

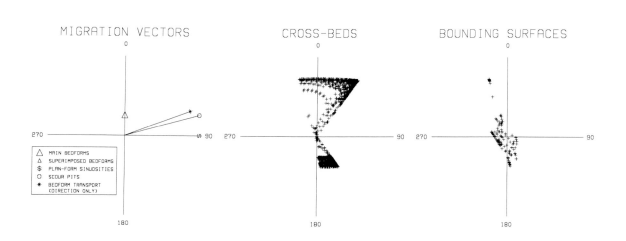

MIGRATION VECTORS

CROSS-BEDS

BOUNDING SURFACES

△ MAIN BEDFORMS
△ SUPERIMPOSED BEDFORMS
$ PLAN-FORM SINUOSITIES
○ SCOUR PITS
✳ BEDFORM TRANSPORT
  (DIRECTION ONLY)

A

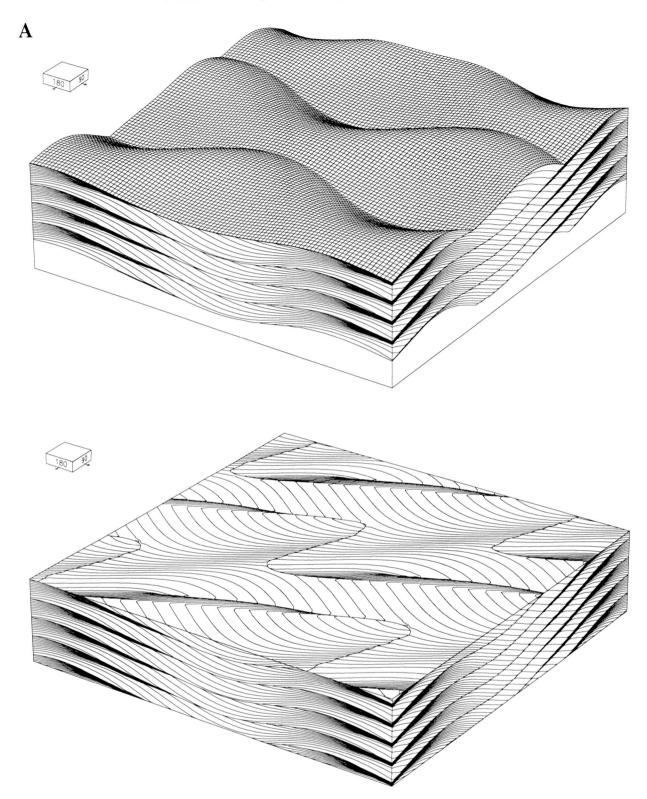

FIG. 43.— Structures formed by bedforms with along-crest-migrating, out-of-phase sinuosities. (A) Rate of along-crest migration of sinuosities is equal to the rate of lateral migration of the main bedform.

(B) Rate of along-crest migration of sinuosities is approximately four times the rate of lateral migration of the main bedform.

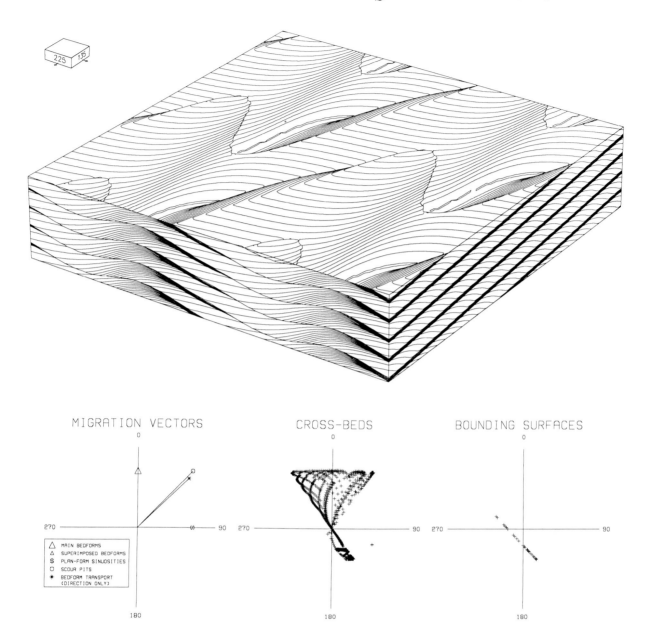

RECOGNITION: In some vertical sections, these structures are relatively similar to structures produced by bedforms with in-phase crestlines (Fig. 42), but the two kinds of structures are distinguishable in horizontal sections.

ORIGIN: These types of structures require conditions that produce oblique bedforms with sinuous out-of-phase crestlines; specific fluid dynamic requirements are unknown.

B

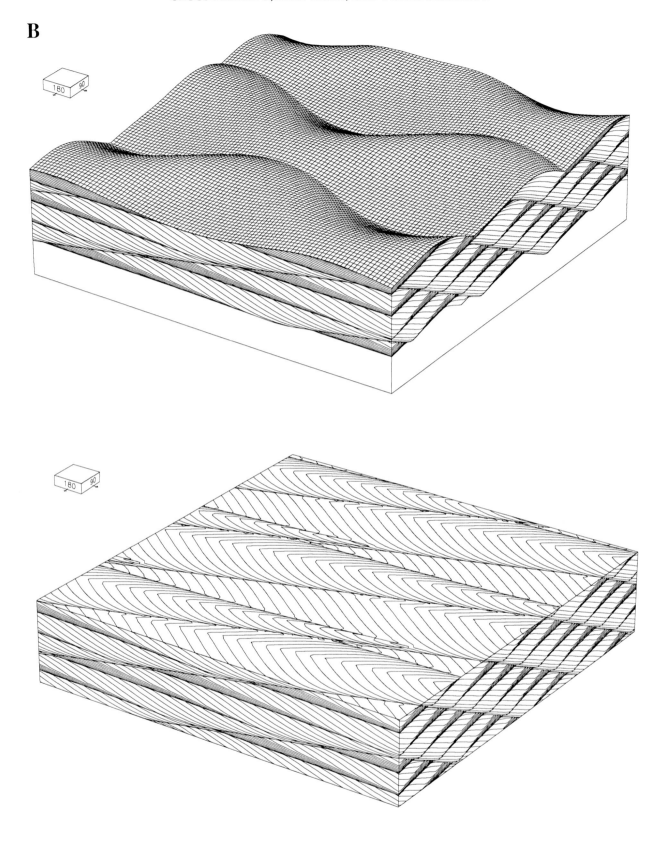

FIG. 43.—*Continued*

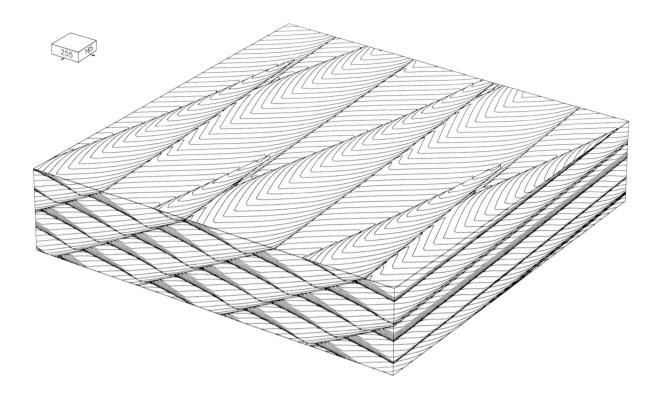

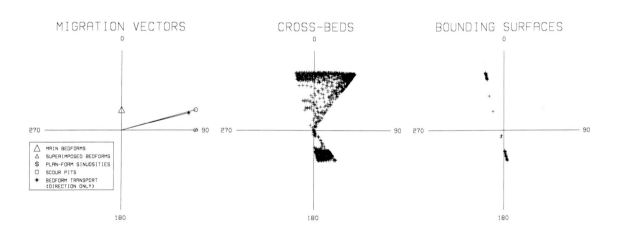

MIGRATION VECTORS

CROSS-BEDS

BOUNDING SURFACES

△ MAIN BEDFORMS
△ SUPERIMPOSED BEDFORMS
$ PLAN-FORM SINUOSITIES
○ SCOUR PITS
✳ BEDFORM TRANSPORT
  (DIRECTION ONLY)

87

FIG. 44.— Structure formed by a dune with a sinuous lee slope but without scour pits in the trough; Navajo Sandstone (Upper Triassic? and Jurassic), Zion National Park, Utah.

RECOGNITION: Bedforms with sinuous crestlines commonly have scour pits in their troughs, but this example demonstrates that such scour pits may be absent. Sinuosity of the lee slope of the dune that deposited this set of cross-beds is demonstrated by the differences in dip direction of the cross-beds in the set. The dip direction is in general toward the viewer but was measured to have a spread of roughly 90°. The concave-up beds in the center of the photograph were deposited in a lee-slope concavity that migrated toward the viewer; the convex-up beds were deposited on a migrating convexity. A uniform-elevation trough profile (rather than a trough with scour pits and intervening spurs) is demonstrated by the planar bounding surface at the base of the set. It is not known whether the uniform elevation of the trough resulted from cohesion of sediment exposed in the dune trough or from in-phase crestlines as shown in Figure 42.

While the dune migrated toward the viewer, the lee-slope sinuosities also migrated from left to right through the outcrop plane. This behavior is recognizable from the asymmetry of the cross-bedding (Fig. 42A). Many of the cross-beds in this structure are themselves cross-stratified; these beds were deposited by superimposed bedforms that migrated across the sinuous lee slope of the main dune. Like the larger lee-slope sinuosities, these superimposed bedforms migrated with a preferred left-to-right component through the outcrop plane (demonstrated by the preferred dip direction of cross-beds deposited by the superimposed dunes). Such migration in a preferred direction across an outcrop plane can result from either the bedform being oblique to the sediment transport direction or from the bedform being oblique to the outcrop. In this example, the dune is suspected to have been oblique to the direction of sediment transport, because the outcrop is more similar in appearance to the oblique-bedform simulation (Fig. 42A) than to the transverse-bedform oblique-outcrop simulation (Fig. 32B).

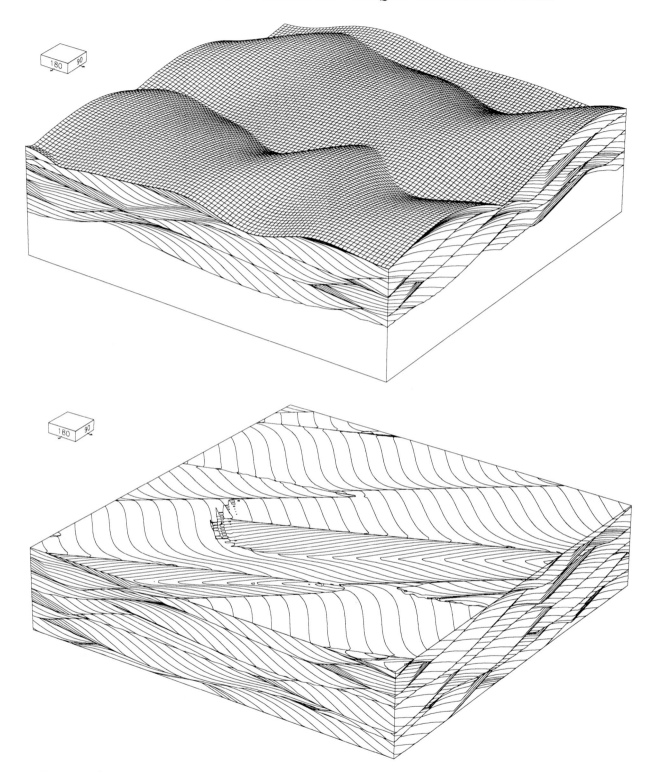

FIG. 45.— Structure formed by oblique bedforms with pseudorandom plan-form geometry.

RECOGNITION: This structure is similar to that shown in Figure 43, but the pseudorandom plan-form geometry adds complexity and causes scatter of the bounding-surface poles. Recognizing that the complexity arises from spatial differences in bedform morphology, rather than from temporal differences in bedform morphology or behavior, requires three-dimensional exposures that include extensive horizontal sections.

ORIGIN: This structure differs from that in Figure 43A only in having a more random plan-form geometry.

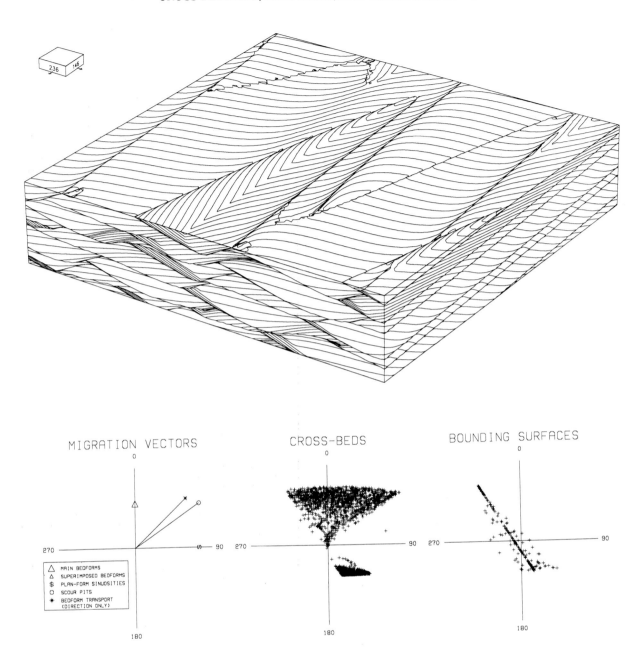

FIG. 45.—*Continued*

FIG. 46.—(Following pages) Structures formed by straight-crested bedforms with scour pits and straight-crested, migrating, superimposed bedforms or lee-side spurs. Real examples of such bedform assemblages are shown in Figures 47-49; real deposits are shown in Figures 50-54.

RECOGNITION: Because of the wide variety of possible sizes, migration speeds, and migration directions of superimposed bedforms relative to the main bedforms, the structures deposited by such bedform assemblages vary greatly in appearance. All of these structures, however, share several features in common. Except where superimposed bedforms have crestlines that exactly parallel the main bedforms, the troughs of the two sets of bedforms intersect to form topographic depressions that geometrically behave like scour pits. Migration of the main bedforms causes the scour pits to migrate with a crest-transverse component (left to right in the computer images); migration of the superimposed bedforms causes the scour pits to migrate with a crest-parallel component of migration (away from the viewer); and deposition causes the scour pits to migrate upward. The resulting scour-pit paths are oblique to the crestlines of the main bedforms. The depositional structures are recognizable in horizontal sections by the oblique orientation of trough axes (scour-pit paths) relative to the trends of the main bedforms. The oblique orientation of the bedforms relative to the transport direction causes a bilateral asymmetry in the distribution of cross-bed dips and bounding-surface dips and causes an asymmetrical distribution of cross-bed dips relative to bounding surface dips.

In horizontal sections, the trend of the main bedforms is indicated by (1) the strike of the relatively continuous foresets deposited where superimposed bedforms did not scour bounding surfaces, either because superimposed bedforms were locally absent or because they climbed at stoss-depositional angles (Figs. 39 and 46D, E, F, and H) and (2) an imaginary line connecting the points at which adjacent scour pits in a bedform trough simultaneously migrated upward through a horizontal section. Note that the main bedform trend does not parallel the bounding surfaces scoured by the superimposed bedforms.

The trend of the superimposed bedforms is not directly observable in either horizontal or vertical sections but is defined by the lines of intersection of the bounding surfaces scoured by the superimposed bedforms and the cross-beds deposited by the superimposed bedforms (Rubin and Hunter, 1983). This principle arises from the fact that during migration of a bedform, the bounding surface being scoured and the cross-bed being deposited intersect along the trough line of the bedform. This trough-line trend (which approximately parallels the crestline trend) can be determined by using either of two techniques. The first technique, which has the most general applicability, uses a stereonet to plot the line of intersection of the cross-bed and bounding-surface planes. A second technique can be employed at those outcrops that fortuitously contain exposures in which the cross-bed traces lie parallel to the bounding-surface traces scoured by the superimposed bedforms. When such outcrops can be located, nature (rather than the stereonet) has performed the appropriate geometric manipulations required to locate the intersecting cross-bed and bounding-surface planes; the trend of the line of intersection of the two planes parallels the outcrop surface. That trend can be measured directly from the outcrop (Fig. 52B).

Structures deposited by bedforms with superimposed bedforms migrating toward a divergent direction are useful for indicating paleotransport directions. Paleotransport directions can be determined relatively precisely if the relative sizes, migration speeds, and migration directions of the different sets of bedforms are known. Where these parameters are not known, as is usually the situation with ancient deposits, these structures are still useful for indicating the quadrant of the paleotransport direction.

ORIGIN: Bedforms with superimposed bedforms migrating in another direction are common in eolian flows (Figs. 47A, 51-53, and 74-76; Rubin and Hunter, 1983), fluvial flows (Fig. 54; Beutner and others, 1967; Boersma and others, 1968), tidal flows (Dalrymple, 1984, fig. 5D; Fig. 47B), and nearshore marine flows (Fig. 50; Rubin, 1987). A wide selection of these structures is illustrated because the structures are varied in appearance, are common, interpretable, and are useful indicators of paleocurrent directions. Flow conditions that are required to produce these kinds of bedform assemblages are discussed in the section *Three-dimensionality caused by superimposed topographic features.*

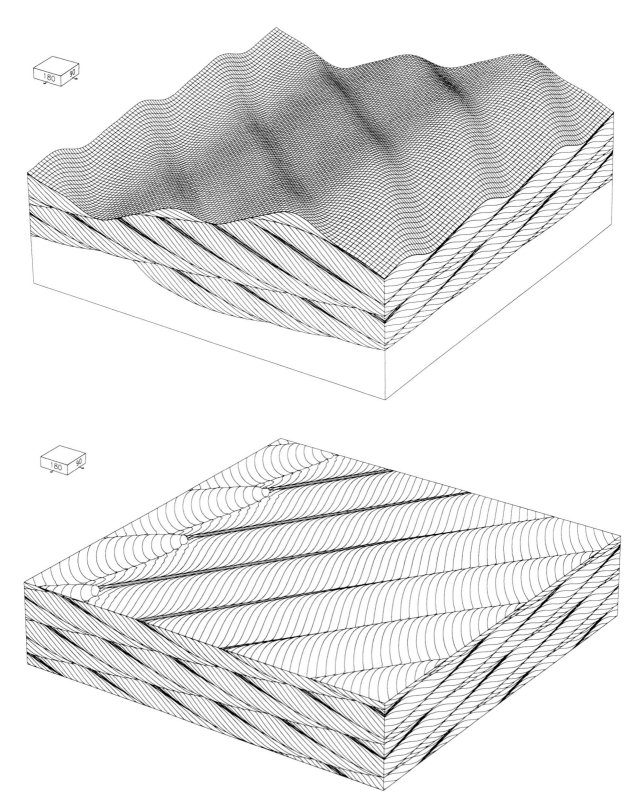

FIG. 46A.— Structures formed by bedforms with superimposed bedforms migrating 45° counterclockwise of the migration direction of the main bedforms. With the exception of the trend and migration direction of the superimposed bedforms, the geometry and behavior of this bedform assemblage is the same as those in Figure 46B and C. Real bedforms with this morphology are shown in Figure 47A.

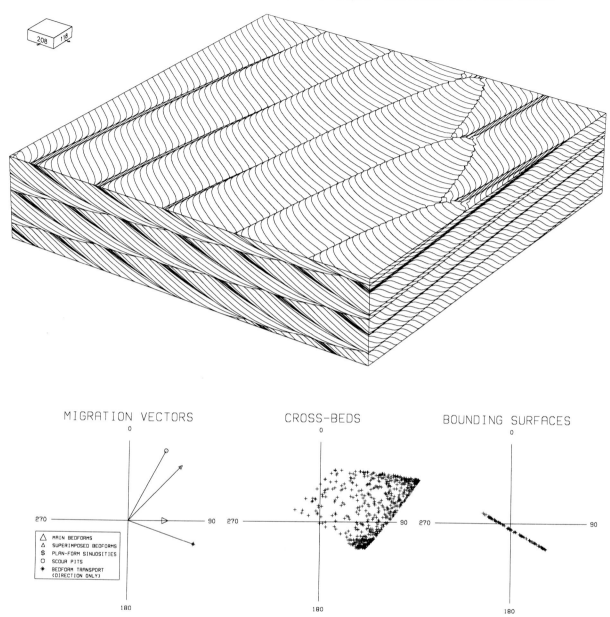

MIGRATION VECTORS

CROSS-BEDS

BOUNDING SURFACES

△	MAIN BEDFORMS
△	SUPERIMPOSED BEDFORMS
$	PLAN-FORM SINUOSITIES
O	SCOUR PITS
✳	BEDFORM TRANSPORT (DIRECTION ONLY)

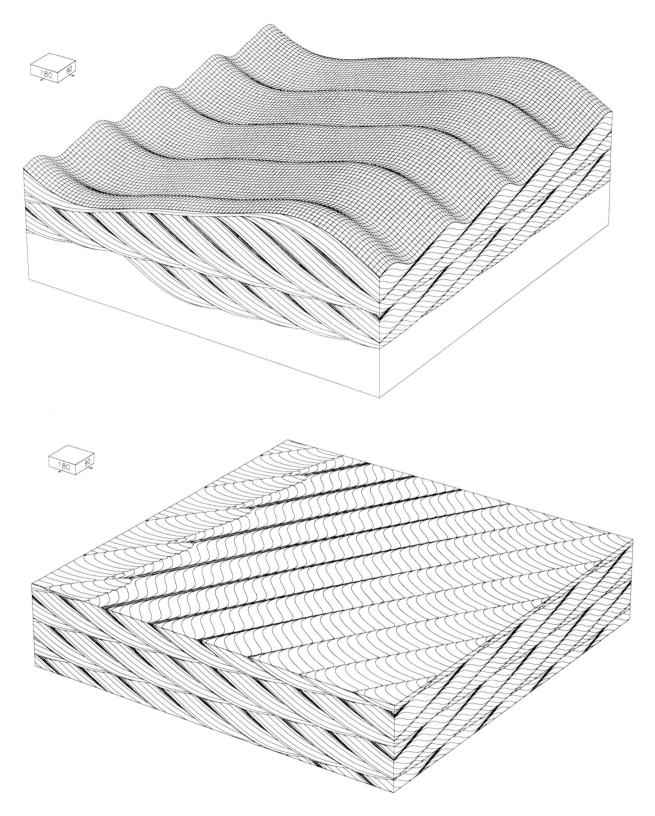

FIG. 46B.— Structures formed by bedforms with along-crest-migrating superimposed bedforms. With the exception of the trend and migration direction of the superimposed bedforms, the geometry and behavior of this bedform assemblage is the same as those in Figure 46A and C.

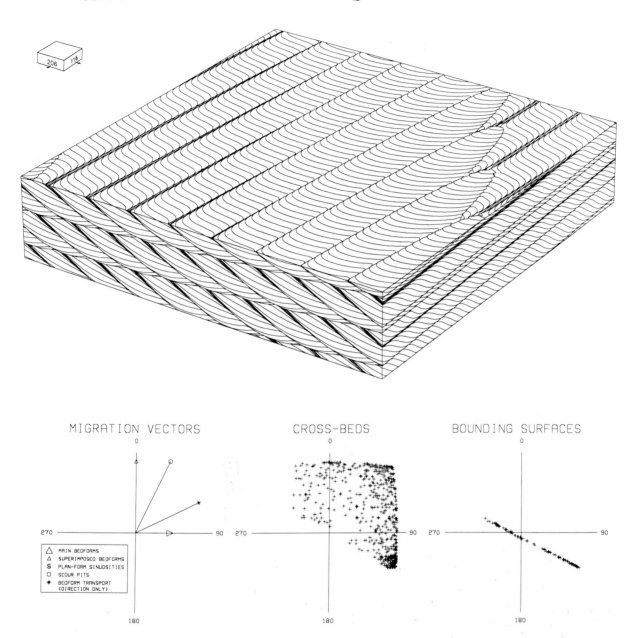

MIGRATION VECTORS

△ MAIN BEDFORMS
△ SUPERIMPOSED BEDFORMS
$ PLAN-FORM SINUOSITIES
○ SCOUR PITS
✳ BEDFORM TRANSPORT
  (DIRECTION ONLY)

CROSS-BEDS

BOUNDING SURFACES

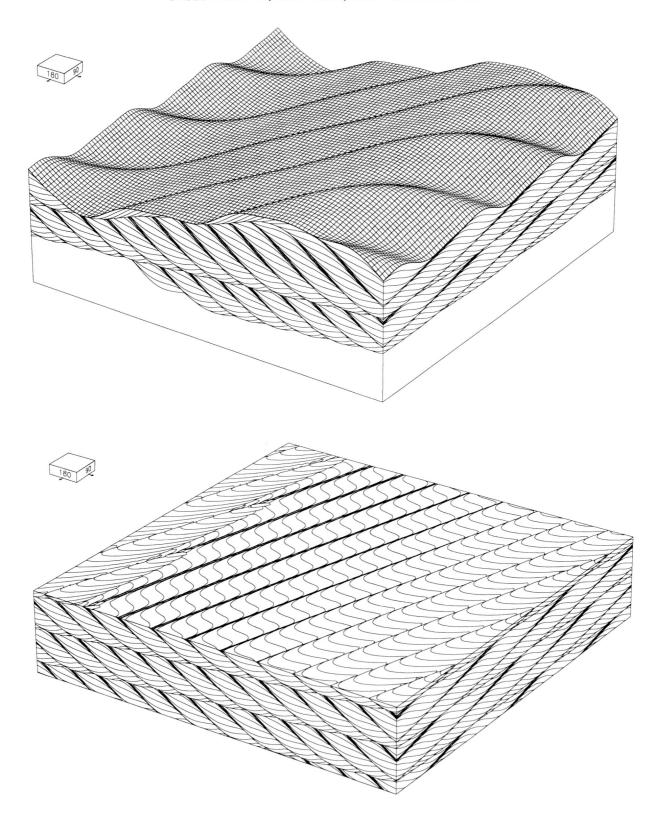

FIG. 46C.— Structures formed by bedforms with superimposed bedforms migrating 135° counterclockwise of the migration direction of the main bedforms. With the exception of the trend and migration direction of the superimposed bedforms, the geometry and behavior of this bedform assemblage is the same as those in Figure 46A and B. An example of this kind of cross-bedding is shown in Figure 53.

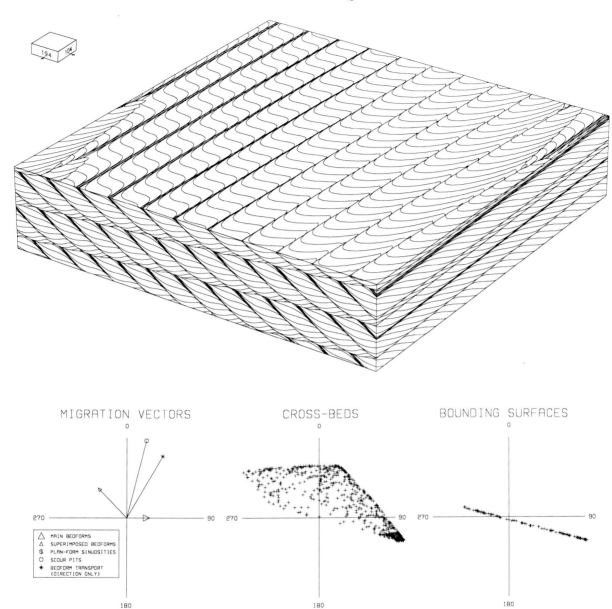

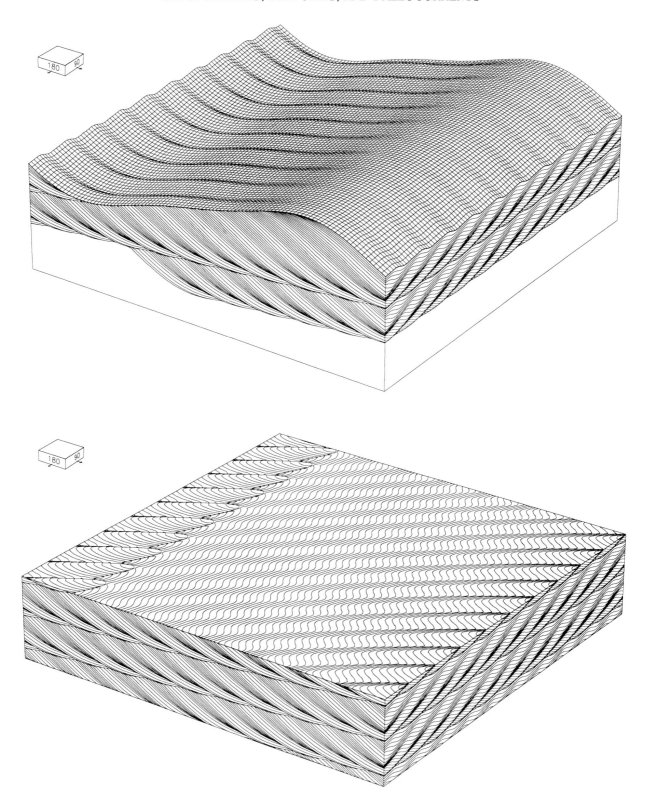

FIG. 46F.— Structure formed by bedforms with along-crest-migrating superimposed bedforms. The bedforms have a mean height ratio of 0.05 and a speed ratio of 1.0. The resulting ratio of along-crest to across-crest transport is 0.05; the transport direction is oriented 86° from the crestline of the main bedform.

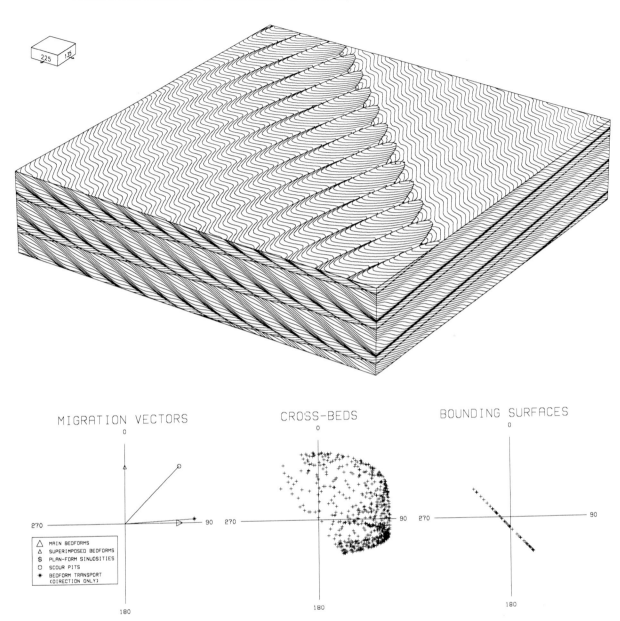

MIGRATION VECTORS

△ MAIN BEDFORMS
△ SUPERIMPOSED BEDFORMS
$ PLAN-FORM SINUOSITIES
○ SCOUR PITS
✳ BEDFORM TRANSPORT
  (DIRECTION ONLY)

CROSS-BEDS

BOUNDING SURFACES

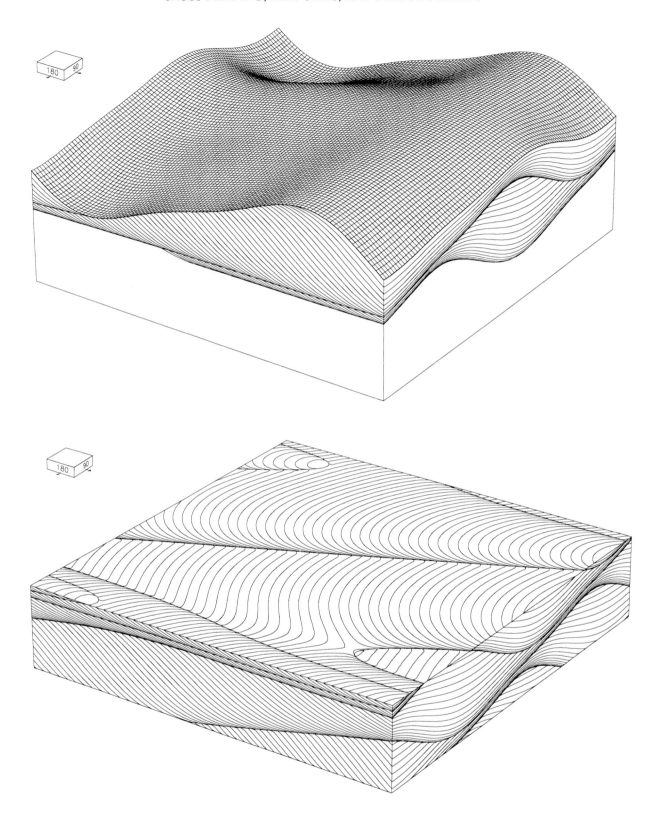

FIG. 46G.— Structure formed by bedforms with along-crest-migrating superimposed bedforms. The bedforms have a height ratio of 0.45 and a speed ratio of 0.3. The resulting ratio of along-crest to across-crest transport is 0.135; the transport direction is oriented 82° from the crestline of the main bedforms.

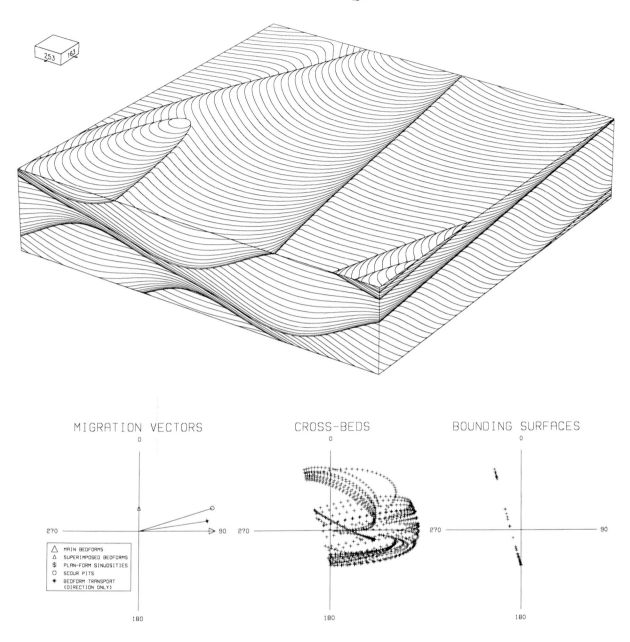

MIGRATION VECTORS

△ MAIN BEDFORMS
△ SUPERIMPOSED BEDFORMS
$ PLAN-FORM SINUOSITIES
○ SCOUR PITS
✳ BEDFORM TRANSPORT
  (DIRECTION ONLY)

CROSS-BEDS

BOUNDING SURFACES

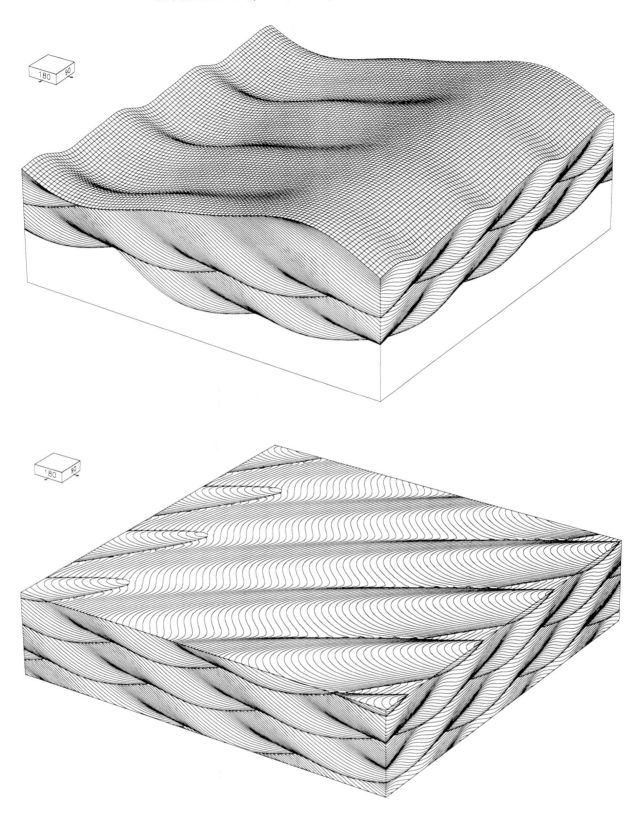

FIG. 46H.— Structure formed by bedforms with along-crest-migrating superimposed bedforms. The bedforms have a height ratio of 0.15 and a speed ratio of 1.0. The resulting ratio of along-crest to across-crest transport is 0.15; the transport direction is oriented 81° from the crestlines of the main bedforms.

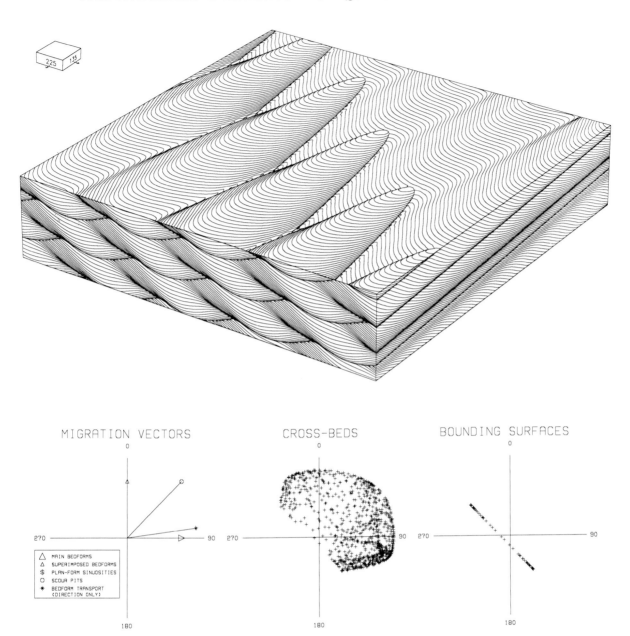

MIGRATION VECTORS

△ MAIN BEDFORMS
△ SUPERIMPOSED BEDFORMS
$ PLAN-FORM SINUOSITIES
○ SCOUR PITS
✳ BEDFORM TRANSPORT
  (DIRECTION ONLY)

CROSS-BEDS

BOUNDING SURFACES

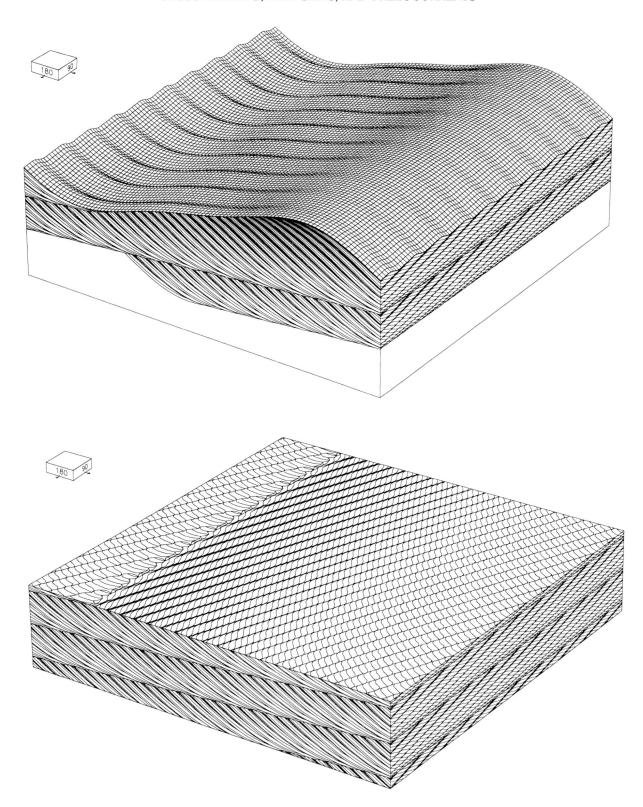

FIG. 46I.— Structure formed by bedforms with along-crest-migrating superimposed bedforms. The bedforms have a height ratio of 0.5 and a speed ratio of 3.0. The resulting ratio of along-crest to across-crest transport is 0.15; the transport direction is oriented 81° from the crestline of the main bedforms.

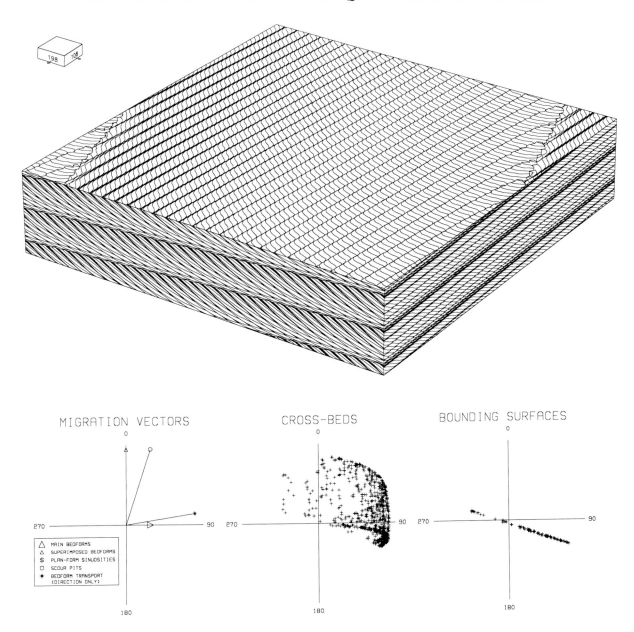

MIGRATION VECTORS

△ MAIN BEDFORMS
△ SUPERIMPOSED BEDFORMS
$ PLAN-FORM SINUOSITIES
○ SCOUR PITS
✳ BEDFORM TRANSPORT
  (DIRECTION ONLY)

CROSS-BEDS

BOUNDING SURFACES

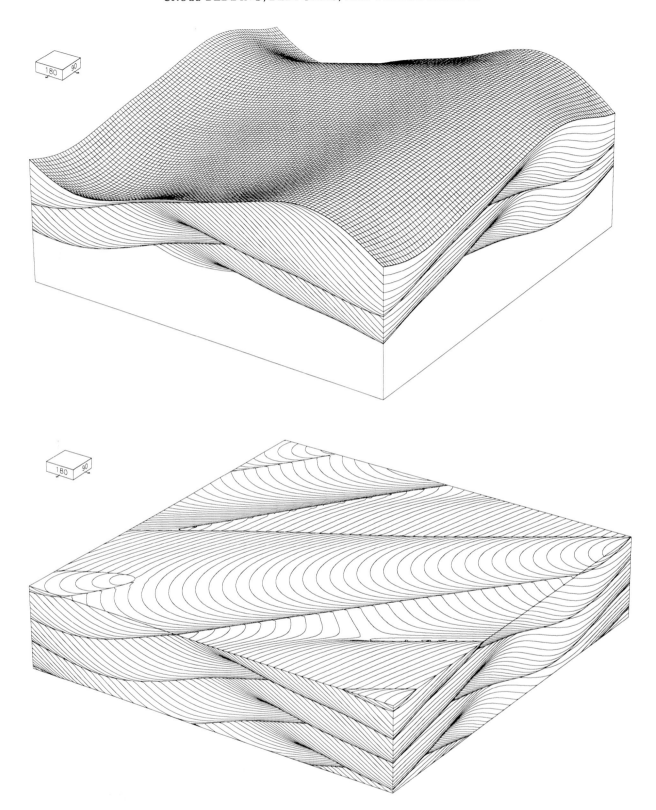

FIG. 46J.— Structure formed by bedforms with along-crest-migrating superimposed bedforms. The bedforms have a height ratio of 0.45 and a speed ratio of 1.0. The resulting ratio of along-crest to across-crest transport is 0.45; the transport direction is oriented 66° from the crestlines of the main bedforms, and they are therefore oblique bedforms.

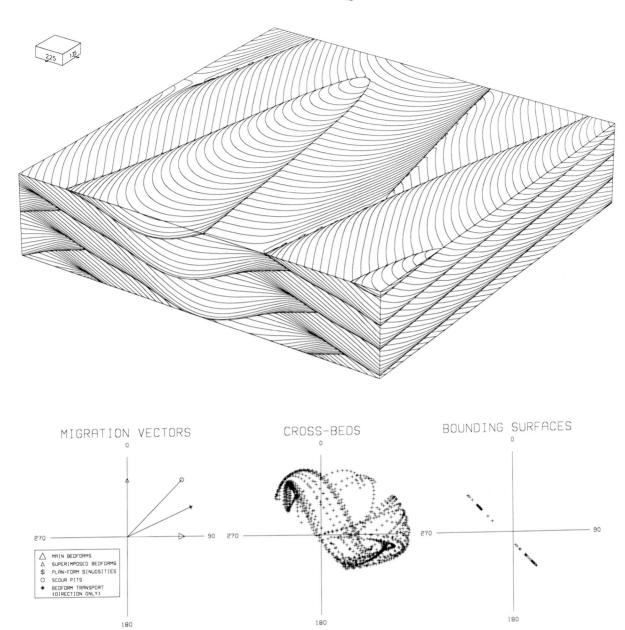

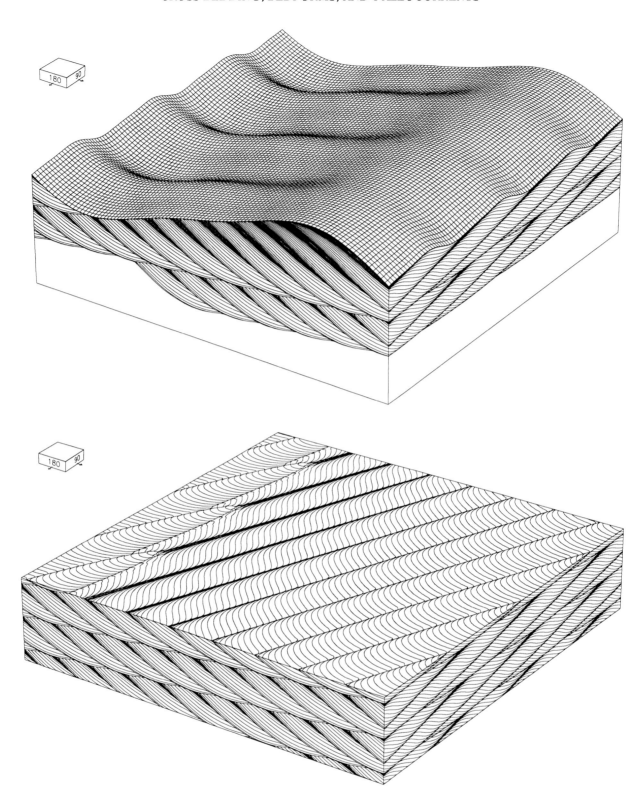

FIG. 46K.— Structure formed by bedforms with along-crest-migrating superimposed bedforms. The bedforms have a height ratio of 0.15 and a speed ratio of 3.0. The resulting ratio of along-crest to across-crest transport is 0.45; the transport direction is oriented 66° from the crestlines of the main bedforms.

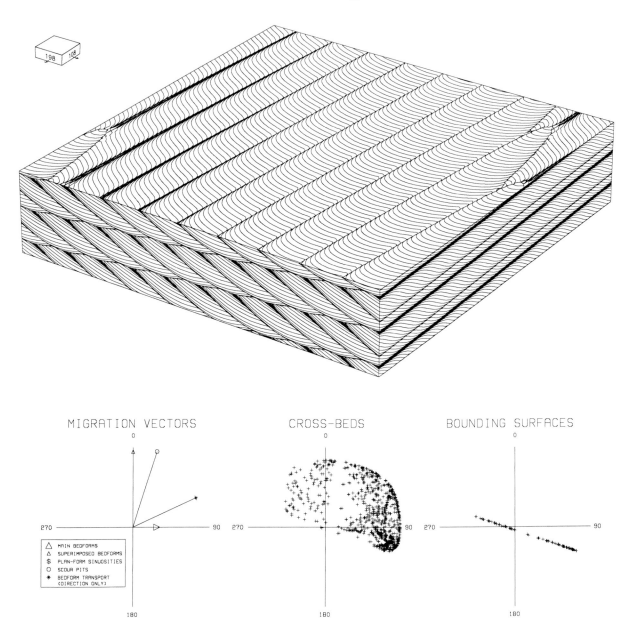

MIGRATION VECTORS

CROSS-BEDS

BOUNDING SURFACES

△	MAIN BEDFORMS
△	SUPERIMPOSED BEDFORMS
$	PLAN-FORM SINUOSITIES
○	SCOUR PITS
✳	BEDFORM TRANSPORT (DIRECTION ONLY)

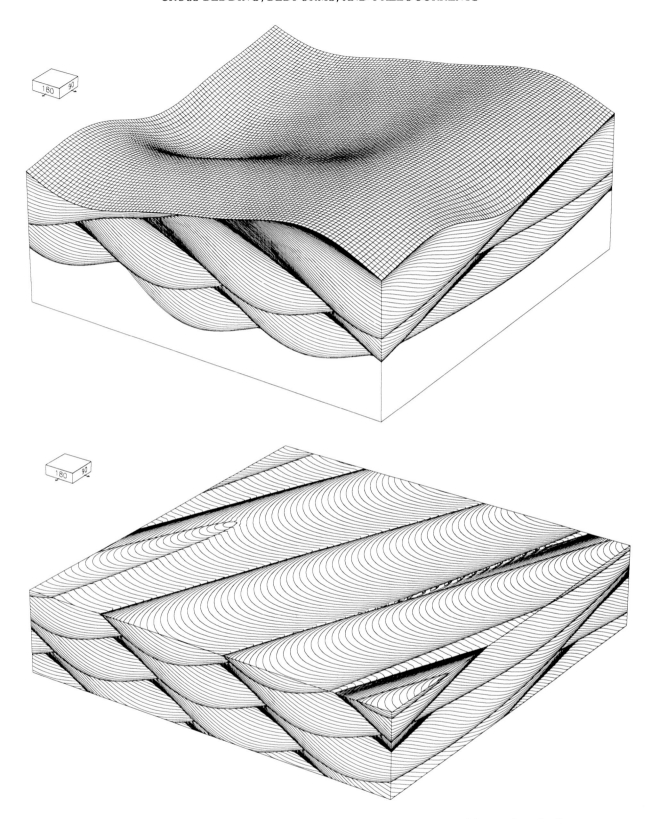

FIG. 46L.— Structure formed by bedforms with along-crest-migrating superimposed bedforms. The bedforms have a height ratio of 0.45 and a speed ratio of 3.0. The resulting ratio of along-crest to across-crest transport is 1.35; the transport direction is oriented 37° from the crestlines of the main bedforms.

114

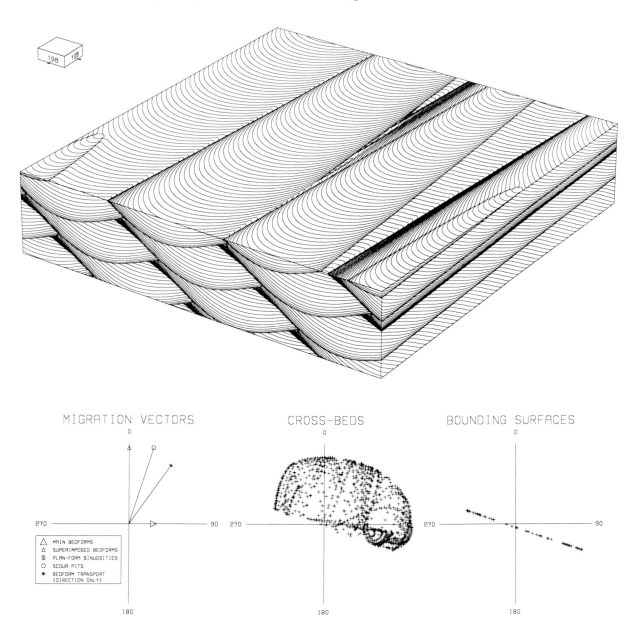

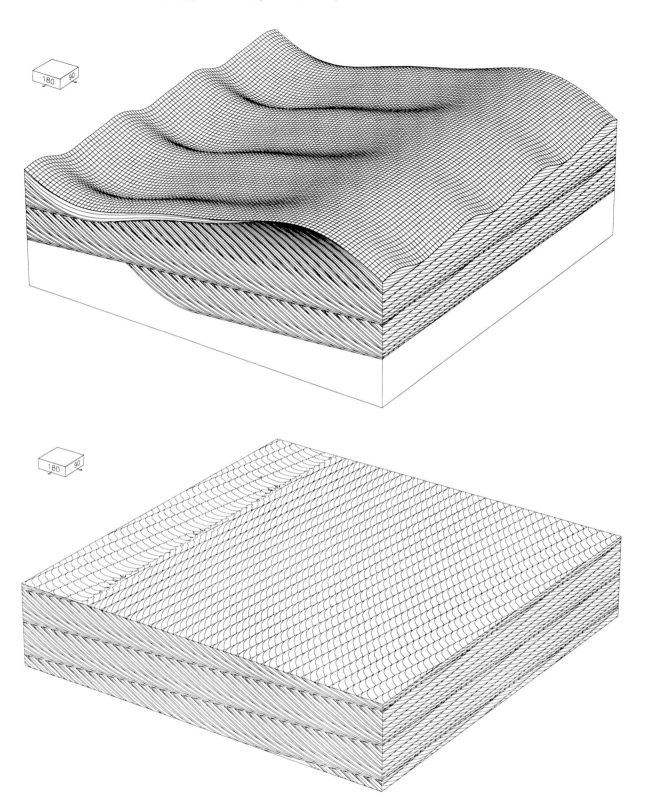

FIG. 46M.— Structure formed by bedforms with along-crest-migrating superimposed bedforms. The bedforms have a height ratio of 0.15 and a speed ratio of 10.0. The resulting ratio of along-crest to across-crest transport is 1.5; the transport direction is oriented 34° from the crestlines of the main bedforms.

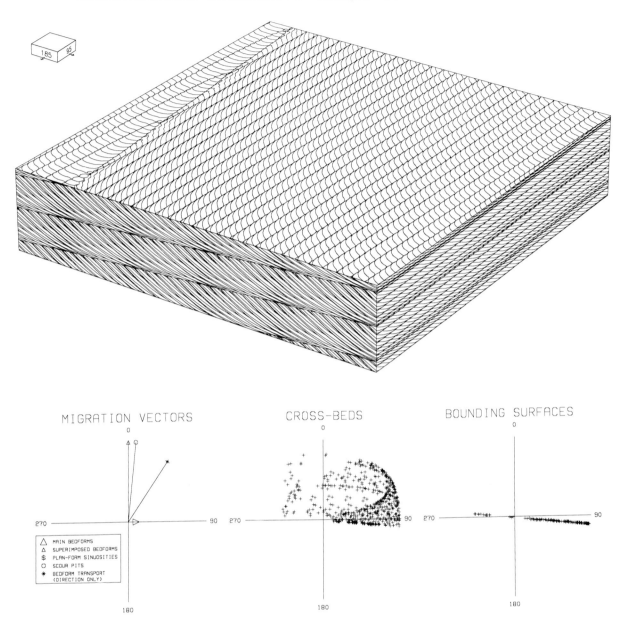

MIGRATION VECTORS

CROSS-BEDS

BOUNDING SURFACES

△ MAIN BEDFORMS
△ SUPERIMPOSED BEDFORMS
$ PLAN-FORM SINUOSITIES
○ SCOUR PITS
✳ BEDFORM TRANSPORT
  (DIRECTION ONLY)

117

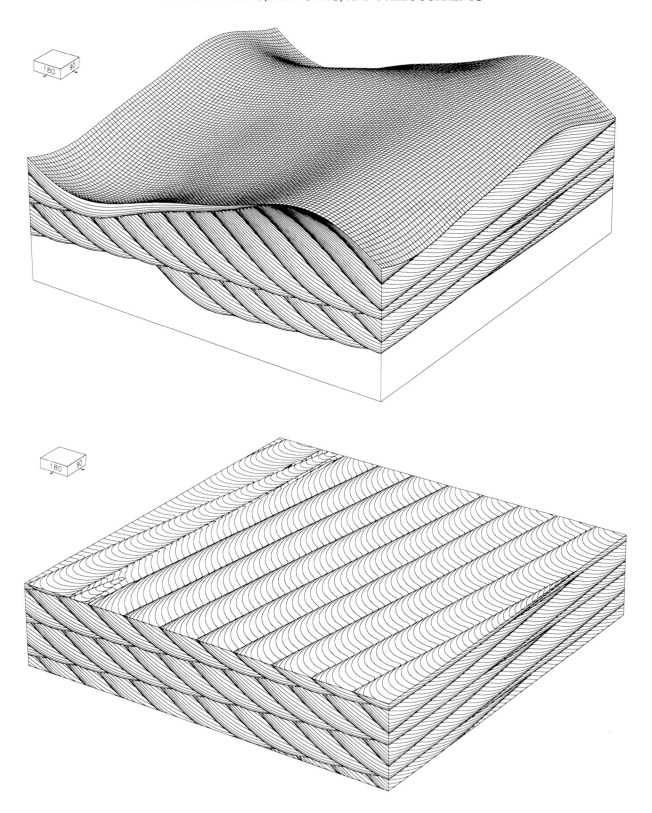

FIG. 46N.— Structure formed by bedforms with along-crest-migrating superimposed bedforms. The bedforms have a height ratio of 0.45 and a speed ratio of 10.0. The resulting ratio of along-crest to across-crest transport is 4.5; the transport direction is oriented 13° from the crestlines of the main bedforms, and they are therefore imperfectly aligned, relatively longitudinal bedforms.

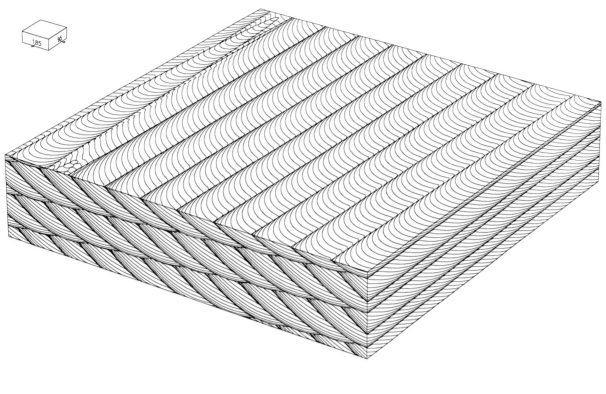

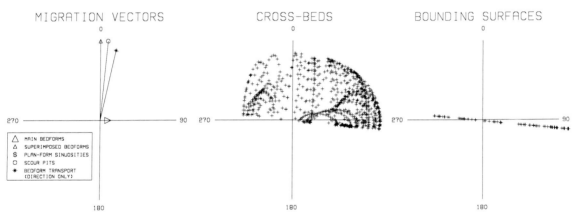

MIGRATION VECTORS

CROSS-BEDS

BOUNDING SURFACES

△ MAIN BEDFORMS
△ SUPERIMPOSED BEDFORMS
$ PLAN-FORM SINUOSITIES
○ SCOUR PITS
✳ BEDFORM TRANSPORT
(DIRECTION ONLY)

119

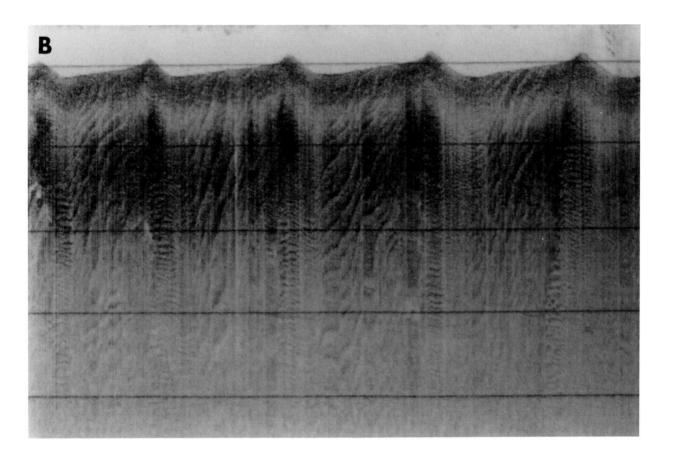

FIG. 47.— Bedforms with superimposed bedforms migrating in a divergent direction. **(A, at left)** Dunes with superimposed dunes, Algodones, California; oblique air photograph taken by Gary Kocurek. The main dune at the lower left is migrating from right to left; the dunes superimposed on its lee slope are migrating toward the viewer, a divergence of 90°, as simulated in Figure 46D-N. At other locations, such as where indicated by the arrow, the divergence is roughly 45°, as simulated in Figure 46A. The height of the main dunes is approximately 10-20 m.

(B, above) Side-scan sonograph of subtidal sand waves with superimposed sand waves, San Francisco Bay, California. The main sand waves are 3-4 m in height and are migrating from left to right. The migration direction of the superimposed sand waves varies with location on the main sand waves; on the dominant lee slopes of the main sand waves, the divergence is roughly 90°. The field of view is 75 m from top to bottom and 600 m from left to right.

FIG. 48.— (Left, above) Oblique dunes in a river meander, Muddy Creek, Wyoming, photographed by Bill Dietrich. The dunes, which are migrating from right to left, become oblique to transport because of cross-stream differences in the rate of advance of the dunes (Dietrich and Smith, 1984). Dietrich (pers. commun.) observed that the spurs migrated up the point bar (away from the viewer) while the main bedforms migrated from right to left. Examples such as this demonstrate that bedforms and their superimposed topographic features can simultaneously migrate in different directions even in steady flows. The channel is approximately 5 m wide, and the largest dunes are 15 cm high.

FIG. 49.— (Left, below) Tidal bedforms with lee-side spurs, Loughor Estuary, South Wales, photographed by Trevor Elliott. Elliott and Gardiner (1981) interpreted these lee-side spurs to be the result of unsteady flow. Helical flow in the lee of the main bedform, inferred to have been caused by a change in flow direction at falling stage, was believed to have created the spurs. Alternatively, flow unsteadiness may not have been necessary to create the bedforms and the spurs (Fig. 48). Trowel is 0.28 m long.

FIG. 50.— (Above) Structure produced by a migrating nearshore bar with along-crest-migrating superimposed bedforms; Pliocene terrace deposits, Monterey Bay, California.

RECOGNITION: This photograph shows details of two scallop-shaped sets of cross-beds in a lateral sequence of approximately a dozen similar scallops (Rubin, 1987). The main bedform that deposited this bed was composed of sand and migrated to the right and toward the viewer. While that sandy bedform was migrating, superimposed gravel bedforms migrated along its trough (toward the left and toward the viewer). Scour pits that were formed by intersections of the troughs of the two sets of bedforms migrated directly out of the plane of the outcrop, toward the viewer. Individual cross-beds within the scallops are curved in plan form and vary in composition. Beds with a dip toward the left are composed primarily of gravel, because they were deposited on the lee slopes of the gravel bedforms migrating along the main trough. Beds which were deposited on the lee slope of the main bedform dip toward the right and are composed mostly of sand. Migration of the main bedform, a nearshore bar, is inferred to have been caused by longshore currents, and simultaneous migration of the smaller bedforms is inferred to have been caused by a rip current.

FIG. 51.— Structures formed by dunes with along-crest-migrating superimposed dunes; Navajo Sandstone (Upper Triassic? and Jurassic), Zion National Park, Utah. The two photographs show different vertical sections through the same beds; arrows (at left) indicate the same set of cross-beds in the two photographs. Note the road for scale in the lower right corner in (B).

RECOGNITION: In (A) the main bedform migrated from right to left, and the superimposed bedforms migrated away from the viewer. The troughs of the two intersecting sets of bedforms formed topographic depressions or scour pits, and migration of each of these scour pits through the plane of the outcrop produced one of the scallop-shaped sets of cross-beds. The sets are shaped like scallops, rather than complete troughs, because each scour pit partially reworked the set deposited by the adjacent scour pit that previously migrated through the outcrop plane. The sets were all truncated on their left sides, rather than randomly truncated, because all the scour pits were migrating down the trough of the main dune, which was migrating toward the left. The origin of this kind of structure is most evident in block diagrams, because those illustrations relate the scour-pit migration paths (in the horizontal sections) to the scallop- and trough-shaped sets visible in vertical sections.

The structure in this outcrop appears most similar to the computer images in Figure 46H, K, and N. In the first of these computer-generated examples (Fig. 46H), the angle between the crestline of the main bedform and the resultant transport direction is $81°$; the bedform is $9°$ from being perfectly transverse to flow. In Figure 46K the angle between the crestline and the transport direction is $66°$ ($24°$ from transverse), and in Figure 46N the angle is $13°$ ($77°$ from transverse). The similar appearance of these differently formed structures demonstrates the difficulty of precise paleocurrent determinations.

(B) shows a second vertical section through the same beds as in (A). (The section in A is viewed from road level slightly off the right side of the photograph shown in B.) The outcrop in (A) cuts across the axes of the scallop-shaped trough sets of cross-beds. In (B), the scallops are nearly undetectable, because the outcrop is nearly parallel to the trough axes; as shown in Figure 46, the scallops are not visible in vertical sections parallel to their axes.

FIG. 52.— Structures formed by a large dune with along-crest-migrating superimposed dunes; eolian deposits in the Entrada Sandstone (Jurassic) near Page, Arizona. (A) shows a horizontal section, and (B) shows a vertical section.

RECOGNITION: The bedding in (A) is not structurally tilted but is a compound set of cross-beds deposited by an assemblage of bedforms like those illustrated in Figure 46I and M. The main dune that deposited this entire coset was at least 30 m high (the thickness of the preserved coset) and migrated from left to right and away from the viewer. The exact migration direction diverges somewhat from the dip direction of the bounding surfaces (Fig. 46). The superimposed dunes were largely preserved because of locally high rates of deposition on the lee side of the main dune. They were several meters high and migrated from the top left of the photograph toward the lower right, along the advancing lee slope of the main dune. Migration of these superimposed dunes scoured the inclined bounding surfaces and deposited the cross-beds that dip to the right. This structure demonstrates the difficulty of distinguishing the deposits of transverse, oblique, and longitudinal bedforms, because it resembles the structures deposited by relatively transverse bedforms (Fig. 46I) and structures deposited by oblique bedforms (Fig. 46M).

In the section shown in (B), the main dune that deposited the entire coset migrated from right to left and toward the viewer. The superimposed dunes migrated into the outcrop, along the lee slope of the main dune. The structure appears to vary from one place to another on the outcrop, but the differences in appearance are due to outcrop curvature rather than real differences in the structure.

At the upper right in (B), the outcrop surface strikes in the same direction as the trend of the superimposed bedforms. Consequently, the cross-bed traces on that part of the outcrop surface parallel the bounding-surface traces, and the cross-bedding appears simple rather than compound (Fig. 46).

FIG. 53.— Structure produced by a dune with super-imposed dunes that migrated obliquely upslope; Navajo Sandstone (Upper Triassic? and Jurassic), Dianah's Throne, near Coral Pink Sand Dunes State Park, Utah. The larger bushes at the upper left of the photograph are approximately 1 m high.

RECOGNITION: The main dune that deposited this coset of cross-beds migrated toward the right and out of the outcrop, roughly in the direction of dip of the bounding surfaces scoured by the superimposed dunes. An upslope component of migration of the superimposed bedforms is evident from the onlapping of the foresets immediately to the left of the center of the photograph, but migration directly up the lee slope of the main bedform can be ruled out at that location because the cross-bed traces are horizontal, whereas the bounding-surface traces are inclined. In contrast, where superimposed bedforms migrate directly upslope, the resulting cross-bed and bounding-surface planes have the same strike, and traces of both planes must be horizontal in the same outcrop plane. Thus, these beds must have been deposited by bedforms that migrated obliquely upslope, as simulated in Figure 46C.

This example is more complicated than the simulation in Figure 46C, because the main dune that deposited the beds in this outcrop was not straight-crested. At the lower left side of the photograph, the dip of simple cross-beds (no compound cross-bedding and therefore no superimposed bedforms) is to the left and toward the viewer; at the right side of the photograph, the dip is to the right and toward the viewer. Unless the superimposed dunes have the same crestline curvature as the main dune, the migration direction of the superimposed dunes relative to the main dune must vary across the outcrop.

128

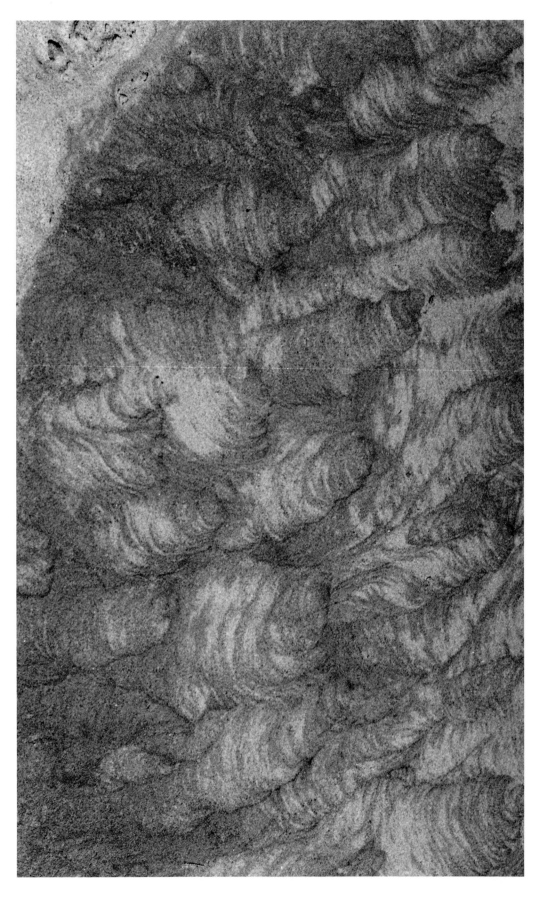

FIG. 54.— Horizontal section through structures deposited by bedforms with migrating lee-side spurs or superimposed bedforms; fluvial deposits, Colorado River, Grand Canyon National Park, Arizona. The area shown is approximately 30 cm from left to right. **RECOGNITION:** The crestlines of the main bedforms (fluvial bars or dunes) that deposited these beds trended from the upper right of the photograph to the bottom center, as indicated by imaginary lines connecting the fingertips of adjacent scour-pit paths. Scour pits are inferred to have been bounded by the main bedforms and by the crests of a set of superimposed bedforms, such as ripples, or bounded by lee-side spurs. Scour-pit paths are controlled by three vectors: migration of the main bedforms (normal to their crestlines in a left-to-right direction), migration of the ripples or spurs along the trough of the main bedforms (bottom to top in the photograph), and deposition (upward through the horizontal section). The scour-pit paths (indicated by axes of the trough-shaped sets, which are seen in horizontal section) are oriented at approximately 45° to the main crestlines, which indicates that the rate of along-crest migration of the scour pits was approximately equal to the migration speed of the main bedforms.

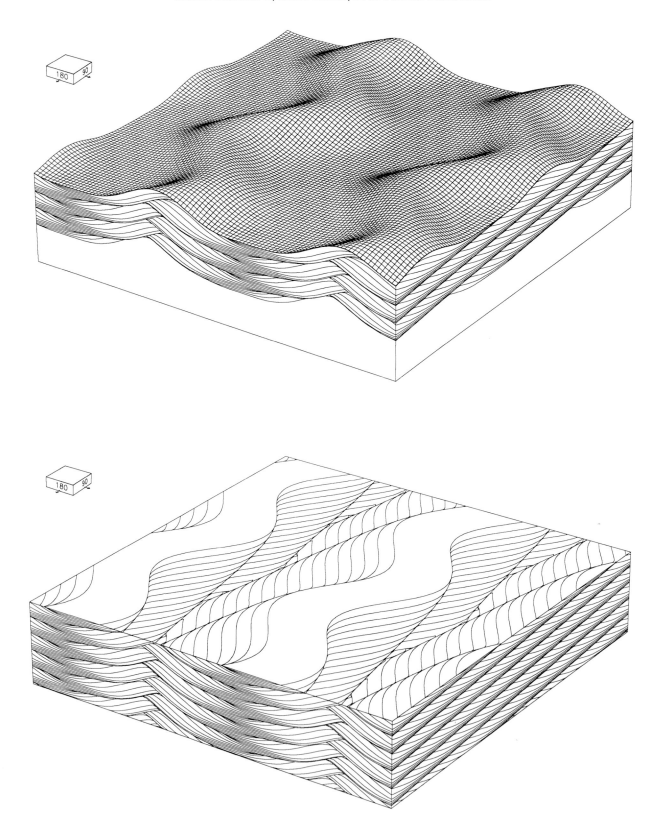

FIG. 55.— Structure produced by perfectly longitudinal (nonmigrating) bedforms with along-crest-migrating sinuosities.

RECOGNITION: In vertical sections perpendicular to the bedform trend, these structures resemble zigzag structures formed by reversing bedforms (Fig. 20)

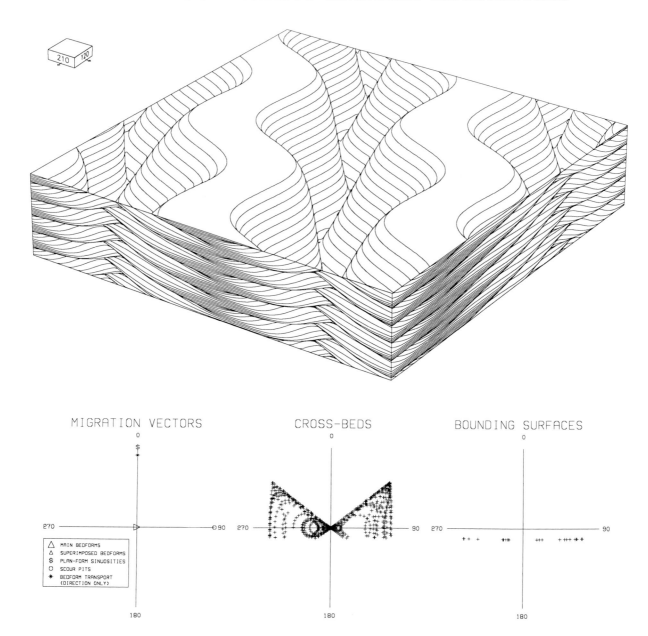

or formed by bedforms with reversing lee-side spurs (Fig. 59). The structures can be distinguished in horizontal sections, because the migrating sinuosities cause cross-beds to dip with an along-crest component.

Cross-beds and bounding surfaces lack dips in the direction of sediment transport (toward 0°). In a natural flow, this gap would be filled in to some extent by the migration of superimposed bedforms (Fig. 56) or by the advancing downcurrent noses of the main bedforms. Because the bedforms that produce this structure do not migrate laterally, the mean cross-bed dip direction varies from location to location.

ORIGIN: Longitudinal eolian dunes with sinuous plan forms were studied by Tsoar (1982, 1983). He documented that the sinuosities migrated along-crest, and he did not detect lateral migration of the dunes. Other examples of this bedform morphology could be expected to include oscillation ripples, tidal sand waves, and tidal ridges. The examples could be expected to be rare, because the reversing flows that produce such bedforms must be exactly balanced to preclude lateral migration of the bedforms. If the individual flow reversals transport too large a volume of sediment relative to the size of the bedforms, then the bedforms will reverse asymmetry, as simulated in Figure 77.

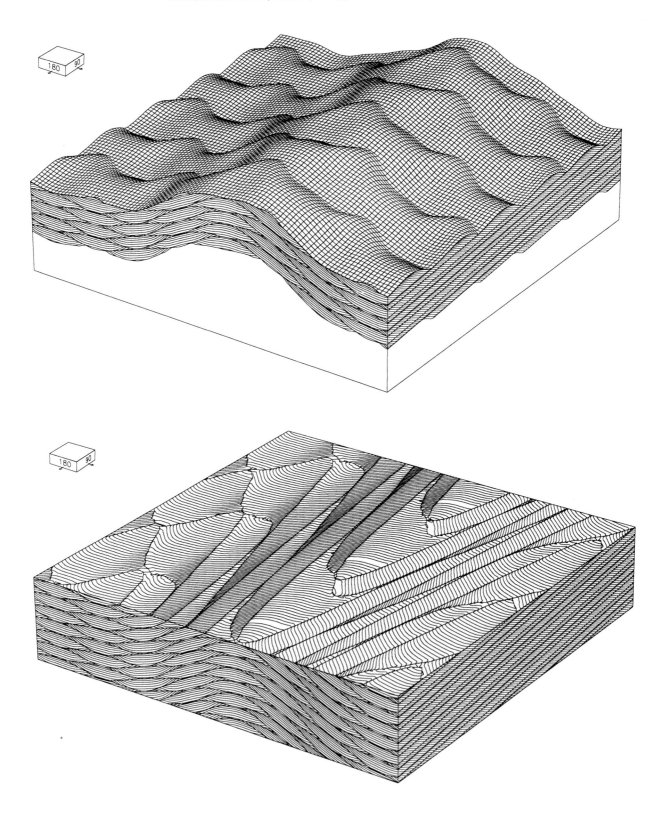

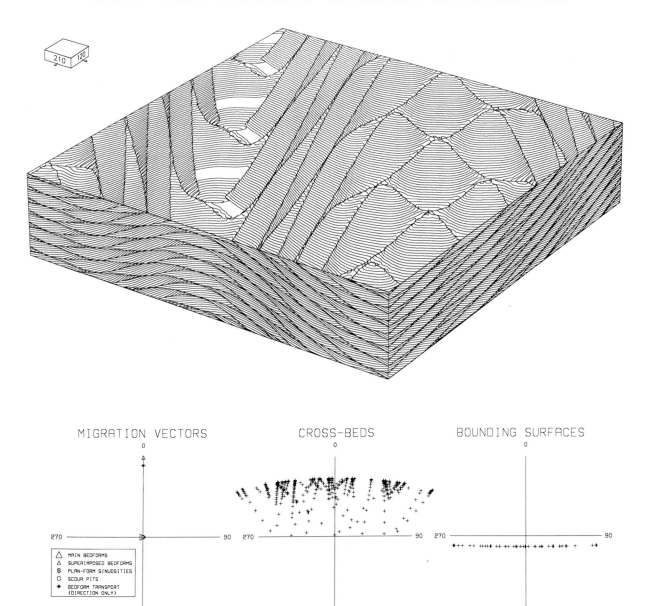

MIGRATION VECTORS — CROSS-BEDS — BOUNDING SURFACES

△ MAIN BEDFORMS
△ SUPERIMPOSED BEDFORMS
$ PLAN-FORM SINUOSITIES
○ SCOUR PITS
✳ BEDFORM TRANSPORT
(DIRECTION ONLY)

FIG. 56.— Structure formed by straight-crested longitudinal bedforms with superimposed, sinuous, out-of-phase transverse bedforms.

RECOGNITION: In sections parallel to the main bedform crestlines, this structure looks like simple climbing-ripple structures, whereas in sections normal to the main bedform crestlines, the structure appears to have been deposited by vertically stacked bedforms.

ORIGIN: As with other longitudinal bedforms, formation of this structure requires the unusual situation where sediment is transported into a depositional area without migration of the bedforms in that area.

Variable Three-Dimensional Bedforms and Cross-Bedding

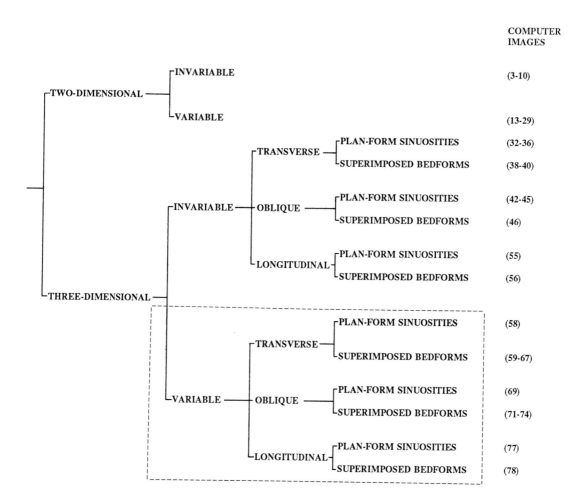

FIG. 57.— Schematic diagram showing the sequence in which illustrations are presented. Dashed line shows which structures are included in the following section.

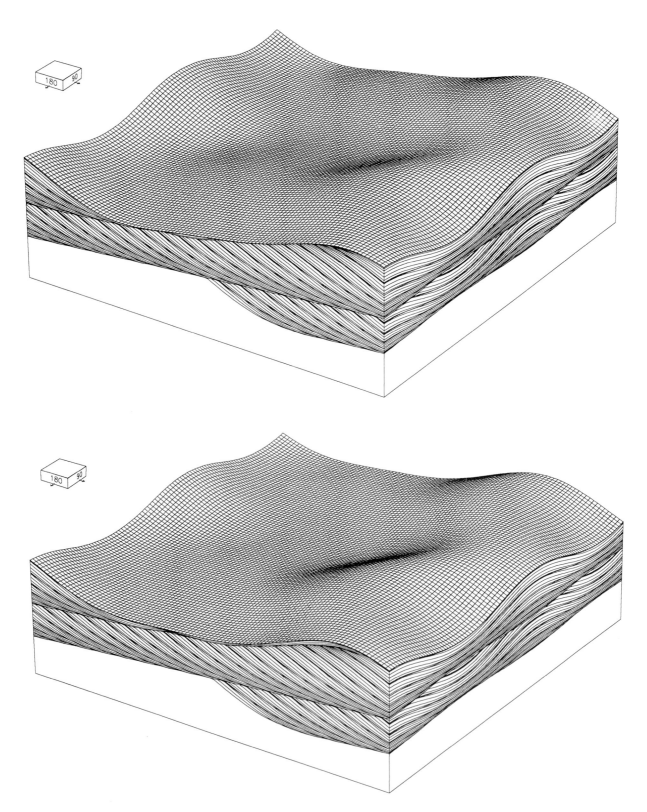

FIG. 58.— Structure formed by sinuous bedforms that fluctuate in migration speed and asymmetry. This depositional situation is virtually identical to the one shown in Figure 22B, except that those bedforms have straight crestlines, whereas the ones shown here have sinuous crestlines. The upper image shows the bedform morphology when asymmetry is a maximum in the asymmetry-fluctuation cycle; the lower image

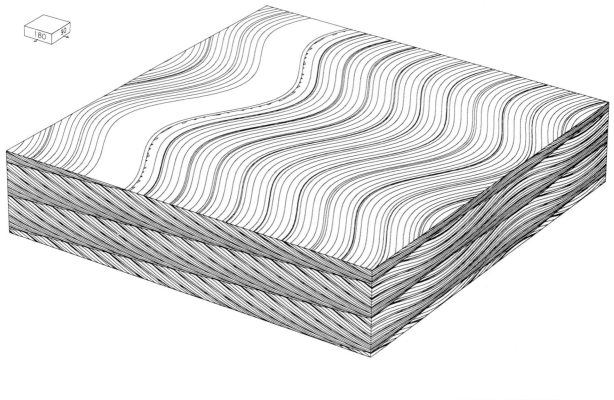

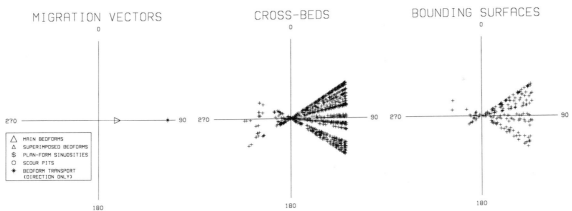

shows the morphology at a later time in the cycle, when bedform asymmetry is a minimum.

RECOGNITION: Although structures deposited by reversing bedforms can closely resemble structures produced by downslope migration of superimposed bedforms that exactly parallel the main bedforms, exact parallelism of the two sets of bedforms becomes less likely as either set of bedforms becomes more three-dimensional. Where the superimposed bedforms do not exactly parallel the main bedform, or where the superimposed bedforms do not have crestlines as long as those of the main bedform, the cross-beds deposited by the superimposed bedforms do not appear conformable—in horizontal sections and in crestline-parallel vertical sections—with the bounding surfaces scoured by the superimposed bedforms (Figs. 65 and 66). In contrast, conformable cross-beds and bounding surfaces (as shown here) suggest a fluctuating-flow origin.

In this structure, and in many of the other structures deposited by variable three-dimensional bedforms, the bounding-surface dip pattern mimics the cross-bed dip pattern. This feature indicates that bounding surfaces have roughly the same shape as the cross-beds.

ORIGIN: This is the same depositional situation as that in Figure 22B, except that these bedforms have sinuous crestlines.

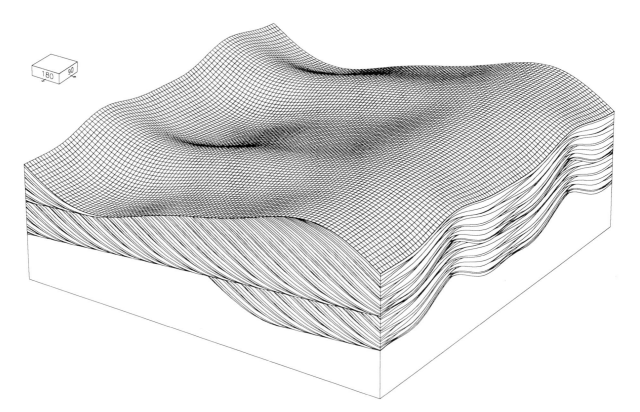

FIG. 59.— Structure formed by migrating bedforms with spurs that reverse asymmetry and migration direction but have no net along-crest displacement. Vertical sections perpendicular to the trend of the main bedforms contain scallops, whereas sections parallel to the main bedforms contain zig-zag structures. Because the zig-zag structures have cross-beds that reverse in their direction of dip in the outcrop plane, the structures might be called herringbone cross-beds by some workers. In a strict sense, however, the structures are not true herringbones because the dip directions are not diametrically opposed.

RECOGNITION: In sections parallel to the trend of the main bedforms, the bedding appears similar to that produced by reversing two-dimensional bedforms (Fig. 18), and in sections perpendicular to the trend of the main bedforms, the bedding is scalloped. The key to identifying this structure is to recognize that the back-and-forth migration of the scour pits and intervening spurs is in a direction that is normal to the migration direction of the main bedform. This behavior is recognizable in horizontal sections or in three-dimensional blocks mapped from vertical sections. In horizontal sections, the trough-shaped sets deposited by the scour pits have a double appearance, most noticeable near the fingertips, where the scour

pits complete their migration upward through the horizontal plane. This double appearance results from reversals in asymmetry of the lee-side system of spurs and scour pits. The deepest point in each scour pit reverses from one side of the scour pit to the other side each time the spurs reverse asymmetry. Before these computer images were generated, these zig-zag structures were thought to form only on the crests of lee-side spurs (Rubin and Hunter, 1983), but the computer images demonstrate that similar structures can form at the bottoms of scour pits. Because scour pits are topographically lower than spur crests, zig-zag structures deposited within scour pits have a higher preservation potential. Where spurs migrate back and forth without changing shape, zig-zags form only at the crests of the spurs.

ORIGIN: This structure is produced by transverse bedforms where the flow direction varies slightly, thereby causing the lee-side spurs to reverse direction of along-crest migration. These structures are useful indicators of paleocurrent direction because they are deposited by transverse bedforms, and, consequently, the trough axes and mean cross-bed dip directions are precise indicators of the paleocurrent direction. Real examples are shown in Figures 60-62.

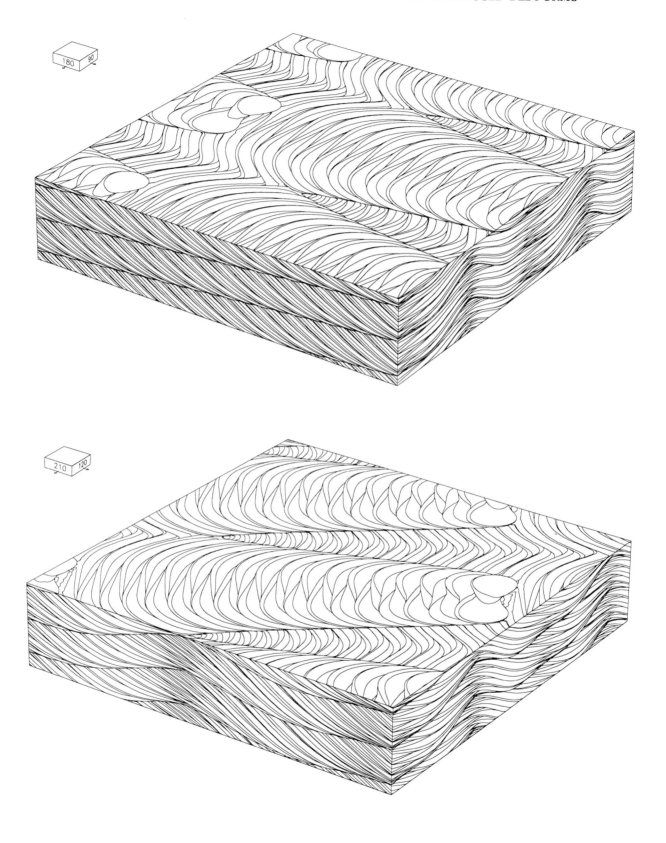

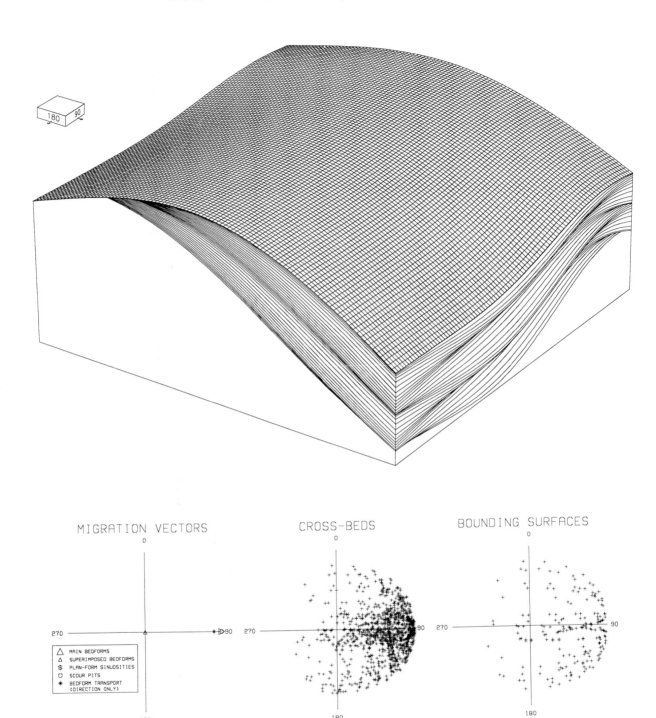

FIG. 59.—*Continued*

FIG. 60.— Zig-zag structure formed by a dune with reversing lee-side spurs and scour pits; Lamb Point Tongue (Upper Triassic?) of the Navajo Sandstone, Kanab Creek, Utah.

RECOGNITION: The back-and-forth migration of the topographic features that deposited this structure is clearly exhibited by the zig-zagging cross-beds and bounding surfaces. Measurements of cross-bed and bounding-surface attitudes showed that the cross-beds dip with a component toward the viewer, the same general dip direction as in simple cross-beds in the area (Hunter and Rubin, 1983). Thus, the spur and scour pits that deposited this structure reversed back and forth across the outcrop plane while migrating toward the viewer; this depositional situation is simulated in Figure 59.

FIG. 61.— Zig-zag structure resembling herringbone cross-bedding; Lamb Point Tongue (Upper Triassic?) of the Navajo Sandstone, Kanab Creek, Utah.

RECOGNITION: The bounding surfaces in this zig-zag structure are inclined at relatively low angles, and, consequently, the outcrop resembles the structures produced within zig-zagging scour pits, as illustrated in the right-hand vertical sections of Figure 59. Although the zig-zags have the appearance of herringbone bedding, herringbone bedding is commonly believed to form by transverse bedforms that reverse their direction of migration frequently, such as with reversing tidal currents. Unless the reversing bed-forms migrate large (and relatively equal) distances during each flow reversal, reversing bedforms will produce structures that are grossly different from herringbone bedding (Figs. 18-21, 63, 64, 69, 77, and 78). Other processes that are likely to produce structures that might be mistaken for true herringbone bedding include: zig-zagging of spurs and scour pits (Figs. 59 and 71), vertical stacking of trough-shaped sets of cross-beds produced by bedforms with out-of-phase scour pits (Fig. 34), and long-term shifts in channel geometry that cause reversals in locations of ebb- and flood-dominated channels.

FIG. 62.— Zig-zag structure deposited by dune with reversing lee-side spurs and scour pits; Navajo Sandstone (Upper Triassic? and Jurassic) near Escalante, Utah.

RECOGNITION: The set of cross-beds in the center of this photograph was deposited by a dune that migrated into the plane of the outcrop. While the dune was migrating, lee-side spurs and scour pits reversed migration direction or asymmetry, thereby creating the zig-zagging sets of cross-beds. Numbers identify the sequence in which specific sets were deposited. Note the alternating dip directions (sets 1 and 3 dip to the left in the outcrop plane; sets 2 and 4 dip to the right). The set of zig-zags is approximately 10 m thick.

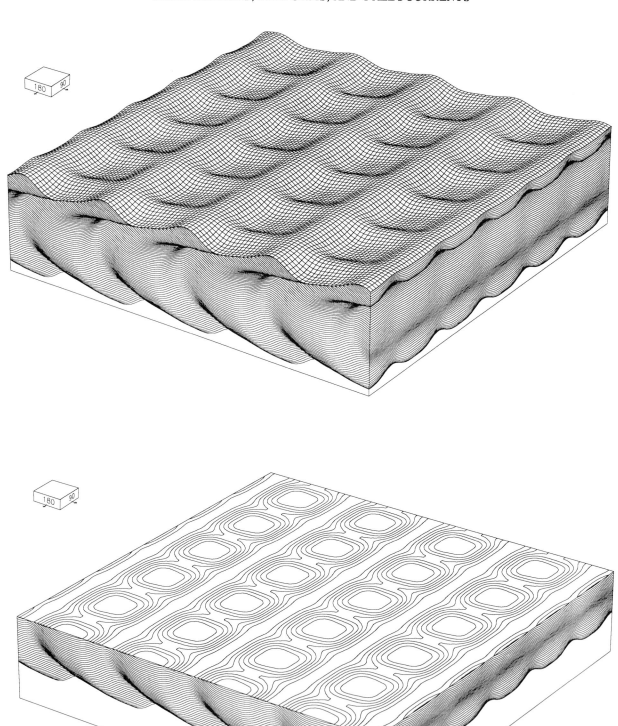

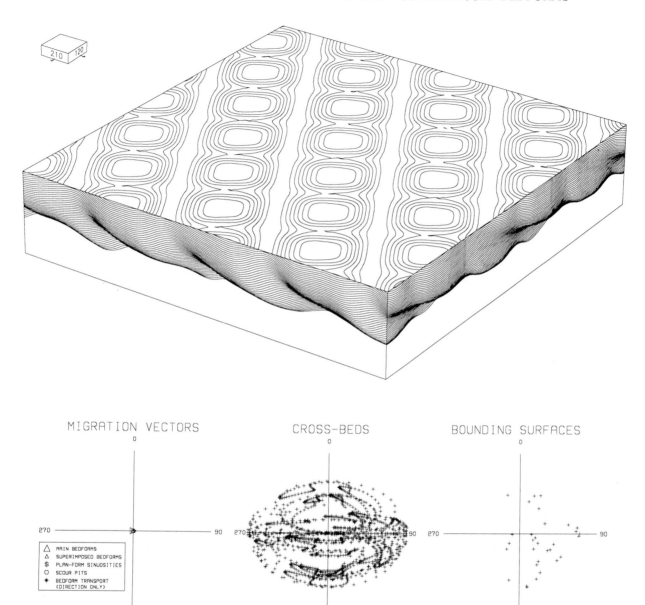

FIG. 63.— Structure formed by reversing, three-dimensional bedforms; one-half of a reversal cycle is shown.

RECOGNITION: The three-dimensional geometry of these bedforms results from superpositioning of features that might be called lee-side spurs, superimposed bedforms, or intersecting bedforms. If produced by waves, this assemblage of bedforms could be called interference ripples, diagonal ripples (Machida and others, 1974), or cross ripples (Clifton, 1976). The three-dimensional structure of these bedforms is evident from the circular foreset traces that are clearly displayed in the horizontal section. Reversals in migration direction of the main bedforms are visible on the vertical section that is transverse to the main crestlines. In this simulation, the small bedforms do not migrate along the troughs of the main bedforms; such migration would cause the structure to resemble the examples shown in Figure 46.

ORIGIN: This bedform morphology is most commonly produced by oscillatory flows, such as wave-generated flows and river eddies with oscillating reattachment points (Fig. 64).

FIG. 64.— Structure formed by reversing ripples with nonmigrating (longitudinal) spurs or superimposed bedforms; fluvial deposits, Colorado River, Grand Canyon National Park, Arizona. This is an example of the structure simulated in Figure 63.

RECOGNITION: In this structure reversals in the direction of ripple migration occurred at three scales, all of which are visible in the vertical section (**A, at left**). At the largest spatial scale and longest temporal scale is a reversal in migration direction throughout the entire bed. Migration directions are right to left at the bottom of the bed and left to right at the top. At a smaller spatial and shorter temporal scale are the several-centimeter-displacement back-and-forth oscillations of the ripples, most clearly visible in the center of the bed, where the angle of climb was vertical. At the smallest and shortest scale, the reversals in migration direction are represented by lamina-to-lamina zig-zags at the ripple crests.

This bed was deposited during high discharge in 1983. Initial deposition was during downstream flow, but as flow receded and a topographic obstruction emerged upstream from the depositional site, an eddy formed, and local flow was directed upstream. Thus, the largest reversal represents one depositional episode (a flood). Smaller reversals may have been caused by the passage of small eddies in the flow or by instabilities in the main eddy.

After the vertical section was photographed, a horizontal section was excavated through the same beds (**B, above**). The horizontal section is at a stratigraphic horizon slightly above where the angle of climb was vertical. The section shows circular cross-bed traces that were deposited by stoss-depositional climb of the scour pits in the bedform troughs.

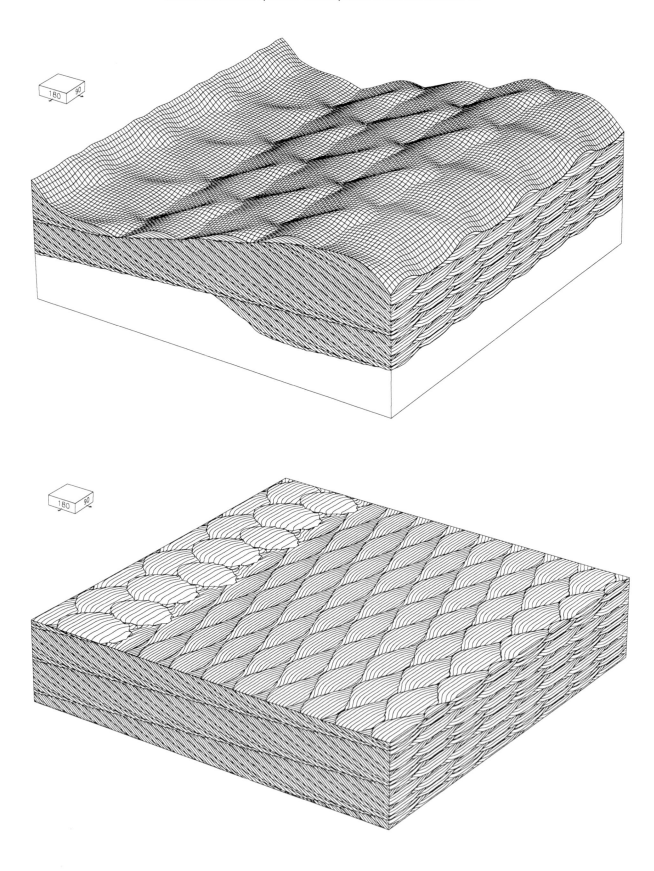

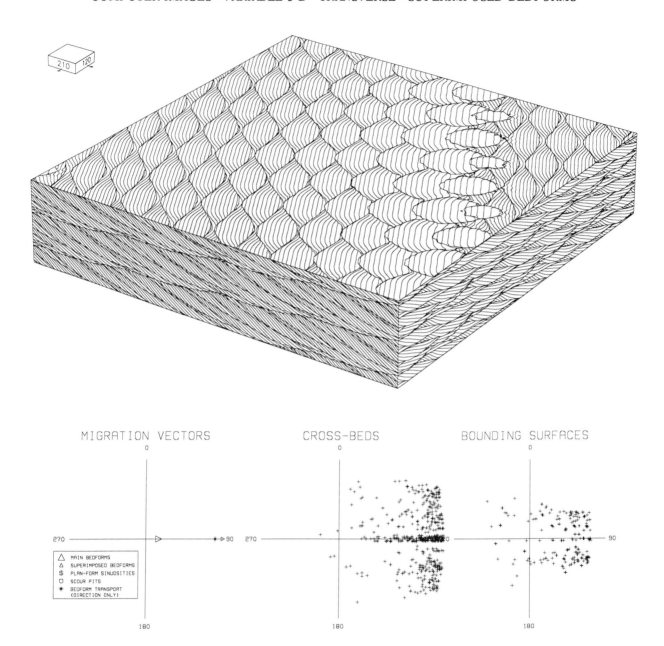

FIG. 65.— Structure formed by two-dimensional bedforms with downslope-migrating superimposed bedforms with sinuous, out-of-phase crestlines.

RECOGNITION: This structure is superficially similar to the structure produced by sinuous out-of-phase bedforms migrating across a horizontal surface (Fig. 34), but two obvious differences arise from the downslope migration of the superimposed bedforms. First, sets of cross-beds deposited by the superimposed bedforms are grouped in larger sets deposited by the main bedforms. Second, downslope migration of the superimposed bedforms causes the bounding surfaces that they scour to dip downcurrent.

ORIGIN: This example is similar to that in Figure 25, but the superimposed bedforms in the example shown here are three-dimensional.

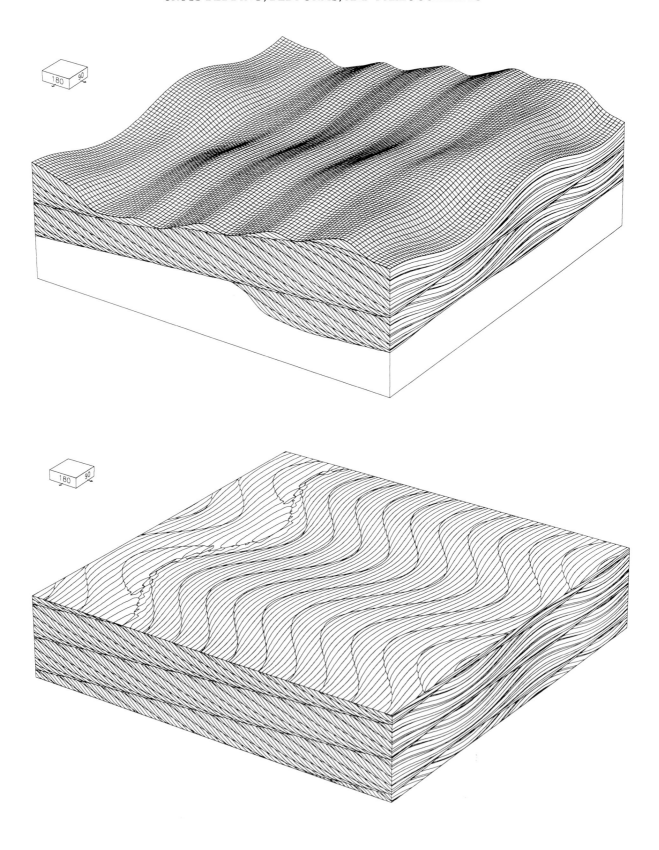

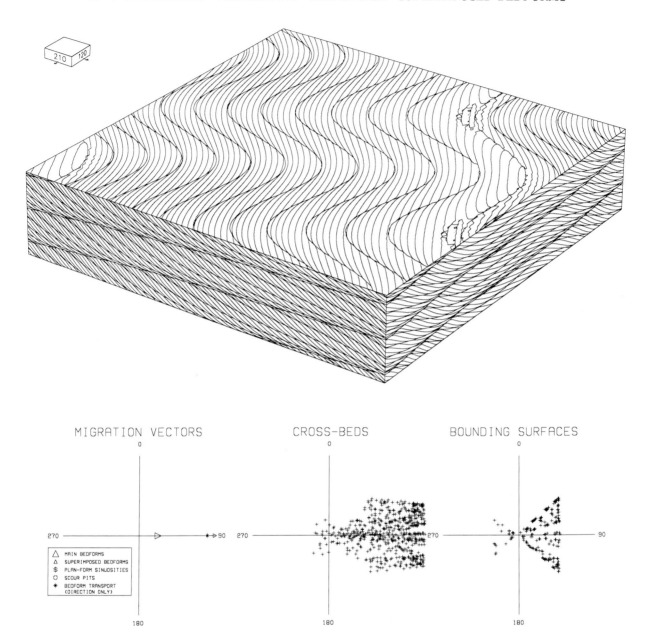

MIGRATION VECTORS CROSS-BEDS BOUNDING SURFACES

△ MAIN BEDFORMS
△ SUPERIMPOSED BEDFORMS
$ PLAN-FORM SINUOSITIES
○ SCOUR PITS
✳ BEDFORM TRANSPORT
(DIRECTION ONLY)

FIG. 66.— Structure formed by two-dimensional bedforms migrating down the lee slopes of three-dimensional bedforms. This depositional situation is identical to the one shown in Figure 25 except that the main bedforms in that example have straight crestlines, whereas the ones shown here have sinuous crestlines. The crests of the superimposed bedforms have the same mean trend as the main bedforms.

RECOGNITION: The crestlines of the superimposed bedforms simulated here do not exactly parallel the crestline of the main bedform. Consequently, the superimposed bedforms do not migrate directly down the main lee slope at all locations. Instead, at most locations the superimposed bedforms migrate obliquely downslope. The along-crest component of the locally oblique migration causes the cross-beds to dip in different directions from the underlying bounding surfaces, as shown in the simpler depositional situation where two-dimensional bedforms migrate obliquely down the lee slopes of larger two-dimensional bedforms (Fig. 46A).

ORIGIN: The origin of this structure is essentially the same as that of the structure in Figure 25, except that here the main bedforms have sinuous crestlines.

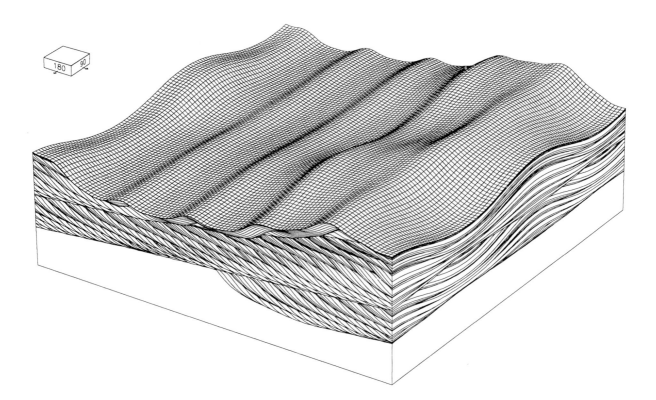

FIG. 67.— Structure formed by reversing, sinuous bedforms with reversing, superimposed, two-dimensional bedforms. This depositional situation combines the bedform reversals in Figure 58 with the superimposed bedforms in Figure 66. The bedforms in the figure above are depicted at a time when they are completing the part of their cycle in which they migrate toward the left; those in the figure on the right are completing their phase of migration toward the right.

RECOGNITION: Interpretation of this structure is extremely difficult, not merely because two processes (reversals of the main bedforms and migration of the superimposed bedforms) form bounding surfaces within sets of cross-beds, but also because the two processes do not have the same period. As a result, the relative phase of the two processes changes through time, thereby causing the subsets of cross-beds to differ slightly from one another. In this structure, the more sinuous bounding surfaces (most clearly distinguished on the horizontal section) were formed by migration of the superimposed bedforms, and the somewhat straighter bounding surfaces were formed by reversals of the main bedform.

Although the superimposed bedforms reverse their direction of migration (simultaneously with the main bedforms), sets of cross-beds deposited by the upslope-migrating superimposed bedforms are not preserved, because the upslope-migrating bedforms always are situated on the eroding side of the main bedform. Real bedforms, however, are not as geometrically perfect as the computer-generated bedforms and may have locations where upcurrent-facing slopes undergo deposition, thereby preserving deposits of upslope-migrating bedforms. Upcurrent-facing slopes can also undergo deposition by systematic processes as well as by random processes. For example, growth of bedforms may involve deposition on upcurrent-facing slopes. In such situations, deposition at the bedform crest can incorporate beds deposited by upslope-migrating bedforms (Dalrymple, 1984).

ORIGIN: This structure requires cyclically reversing flows—to produce cyclically reversing bedforms—and requires flows that maintain superimposed bedforms.

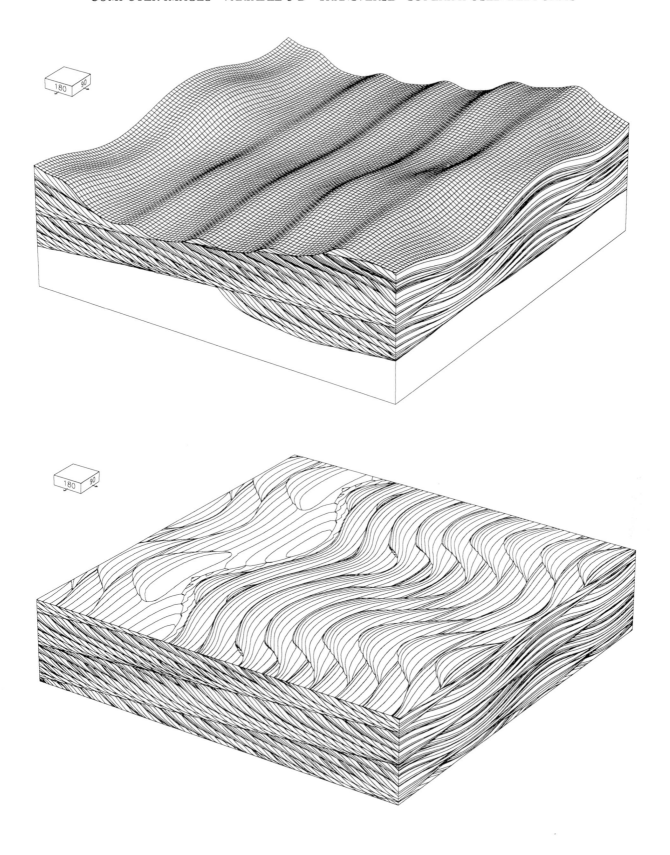

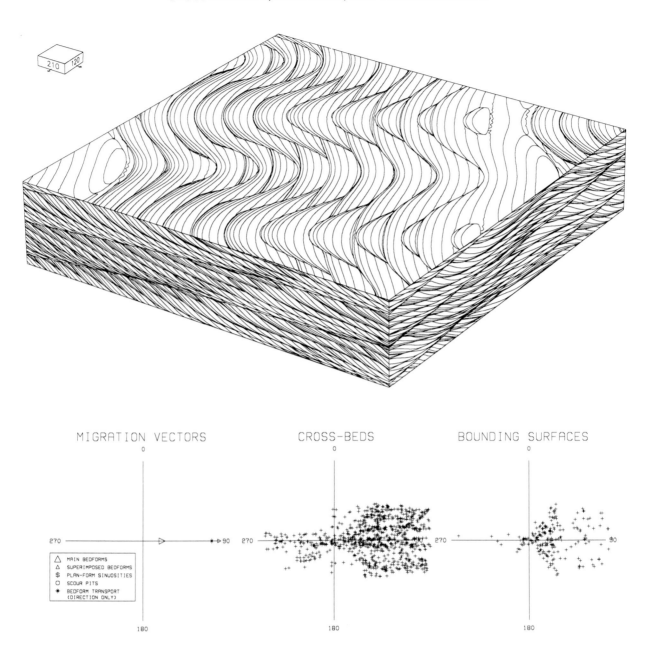

FIG. 67.—*Continued*

FIG. 68.— Structures with compound cross-bedding interpreted to have formed both by fluctuating flow and superimposed bedforms; eolian deposits, Navajo Sandstone (Upper Triassic? and Jurassic), Zion National Park, Utah. Note the person for scale near the lower right corner.

RECOGNITION: The lowest two sets of cross-beds in this photograph have bedding characteristics that are best explained by fluctuating-flow processes like those illustrated in Figures 22 and 58: basal wedges (A) and internal bounding surfaces that are relatively conformable with underlying foresets (B). In contrast, the set of cross-beds that occupies most of the upper half of the photograph contains subsets of cross-beds (C) that were deposited by superimposed bedforms, evident from the differing dip directions of the cross-beds and subset bounding surfaces and from the trough-shaped bounding surfaces of some of the subsets (D).

Although this photograph shows compound cross-bedding formed by reversing of the main bedforms (lower part of photograph) and formed by migration of superimposed bedforms (upper left), the structures are in different sets of cross-beds and were therefore deposited at different times and by different dunes, not simultaneously on the same bedform as is illustrated in Figure 67. The same set that contains the subsets deposited by the superimposed bedforms, however, also has other subsets (E) with relatively conformable set boundaries. These subsets resemble reversing-bedform deposits. If such an interpretation is correct, then superimposed bedforms and fluctuating flow both affected this dune, although not necessarily at the same time.

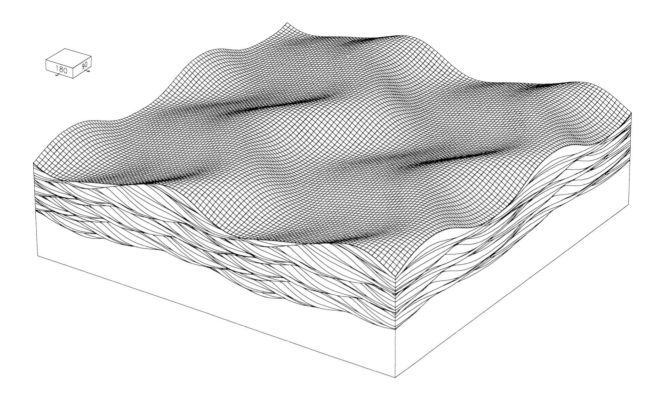

FIG. 69.— Structure formed by migrating, asymmetry-reversing bedforms with along-crest-migrating sinuosities. The migration speed of the bedform is equal to the along-crest speed of the crest-line sinuosities. Bedform morphology is virtually identical to that shown in Figure 55, but in that figure the bedform is nonmigrating (perfectly longitudinal) and nonreversing. The image above shows bedform morphology at a time in the asymmetry cycle when the bedforms face most steeply to the right; the image on the right shows bedform morphology at a later time in the same asymmetry cycle, when the bedforms face most steeply to the left.

RECOGNITION: Along-crest migration of the crest-line sinuosities of these bedforms is obvious in the horizontal sections and is also expressed in the preferred dip (toward the right) in the vertical sections parallel to the bedform crestlines. Examination of horizontal and vertical sections on the block diagram shows that the scallop-shaped bounding surfaces strike in the same direction as the bedform trend, which indicates that the bounding surfaces formed by changes through time in morphology of the entire bedform rather than by along-crest-migration of superimposed features.

ORIGIN: Bedforms capable of producing this kind of structure are probably common in flows that reverse direction, but not by exactly 180°. Likely bedforms include some linear eolian dunes, tidal sand waves and ridges, and oscillation ripples.

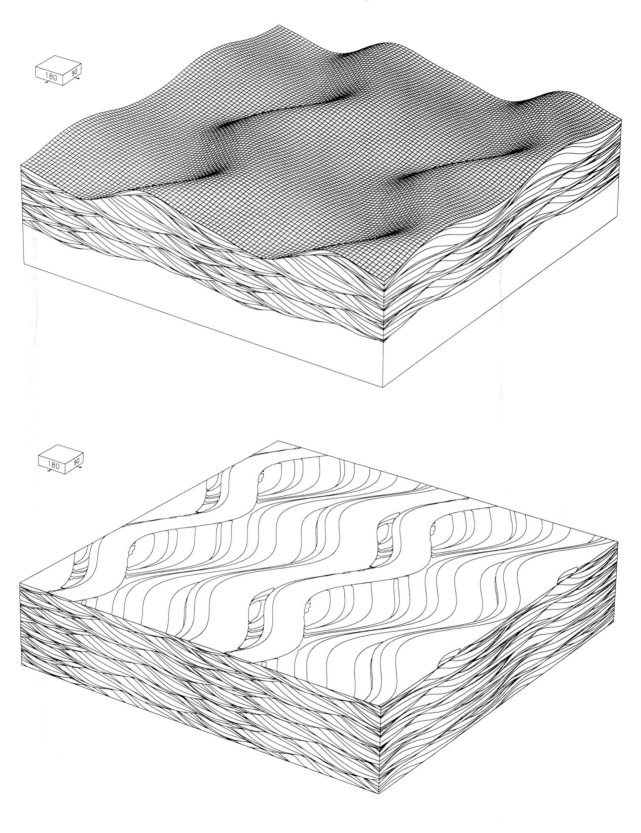

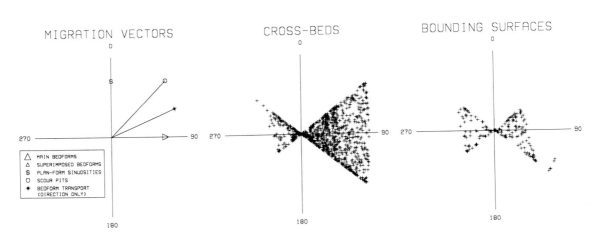

FIG. 69.—*Continued*

FIG. 70.— Migrating, stoss-erosional, lee-depositional, oblique oscillation ripples; Moenkopi Formation (Triassic), Capitol Reef National Park, Utah.

RECOGNITION: An oscillatory-flow origin of these ripples is suggested by ripple symmetry and by the relatively two-dimensional morphology of the ripples. Migration of the ripples toward the upper right of the photograph was accompanied by erosion on the ripple stoss sides (lower left) and deposition on the lee sides (upper right); erosion on the stoss sides has exposed beds that were deposited on the ripple lee sides when the ripples were positioned to the lower left. Erosion of the ripple flanks was more severe on ripple flanks facing toward the left than on flanks facing toward the bottom of the photograph. This feature suggests that the crestline sinuosities were migrating with a longitudinal component of transport (from left to right along the ripple crestlines), as simulated in Figure 69. The resultant transport direction thus is oblique to the trend of the ripples, in the general direction indicated by the arrow. If the individual oscillations in the flow that created these bedforms transported enough sediment to cause reversals in ripple asymmetry, then Figure 69 may be a realistic simulation; if the individual flow oscillations were too brief or too weak to transport enough sediment to effect ripple asymmetry, then the ripples may have behaved as invariable bedforms.

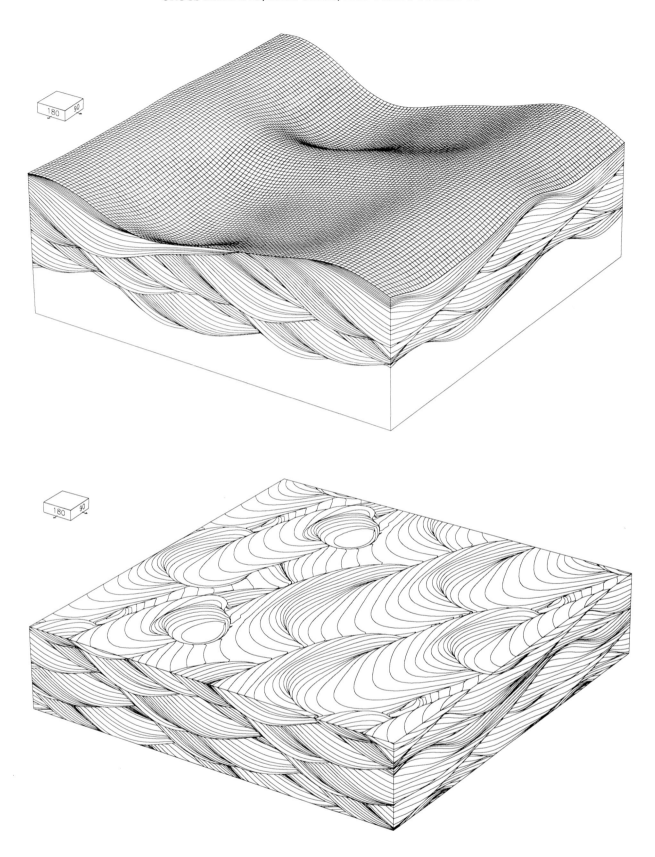

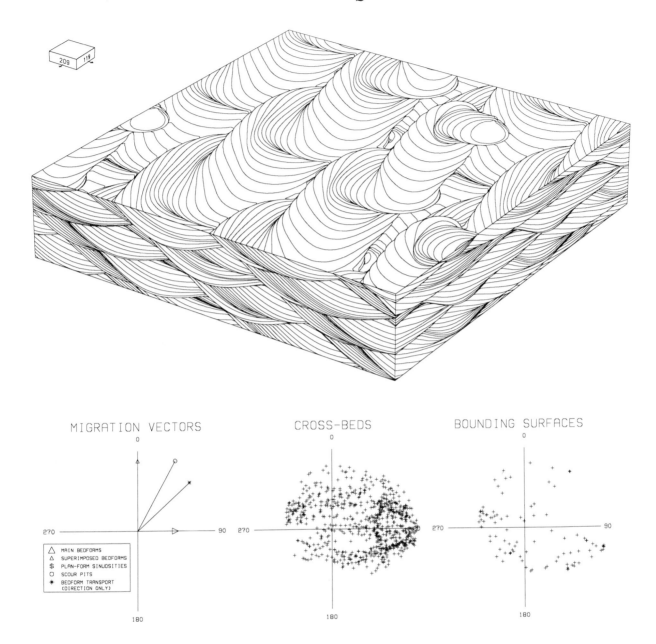

FIG. 71.— Structure formed by migrating bedforms with spurs that reverse asymmetry and migration direction and have net along-crest migration. This structure is an oblique-bedform analog of the structure shown in Figure 59 and differs from it in that the spurs have a net along-crest migration.

RECOGNITION: The main bedform in this example is oblique to transport. The lee-side scour pits and spurs migrate along-crest in addition to reversing back and forth. This distinctive scour-pit behavior causes scour pits to follow zig-zagging paths that produce zig-zagging, trough-shaped sets of cross-beds (most clearly recognized in horizontal sections). As with other deposits of unsteady three-dimensional bedforms, the complex bed morphology and behavior cause bounding surfaces to dip with a wide range of azimuths and inclinations. Consequently, dips of the bounding surfaces plot as scatter diagrams. Along-crest migration of the spurs prevents the formation of vertically zig-zagging structures, as shown in Figure 59.

ORIGIN: The origin of this structure is similar to that shown in Figure 59 except that those bedforms are transverse to transport and have spurs that do not migrate along-crest; the origin is also similar to that of some examples in Figure 46, but those bedforms have scour pits that migrate in straight lines, and the scour pits shown here migrate with zig-zagging paths.

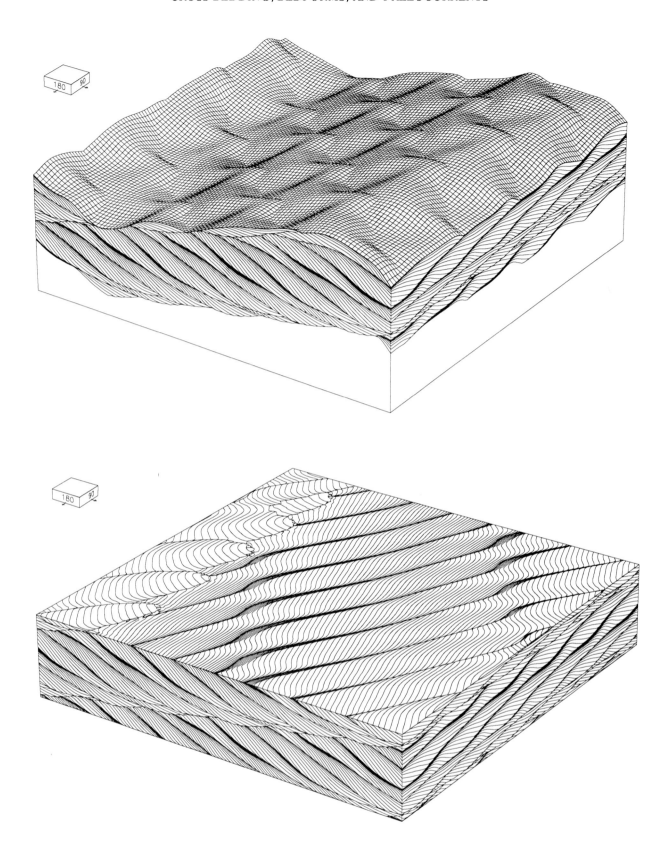

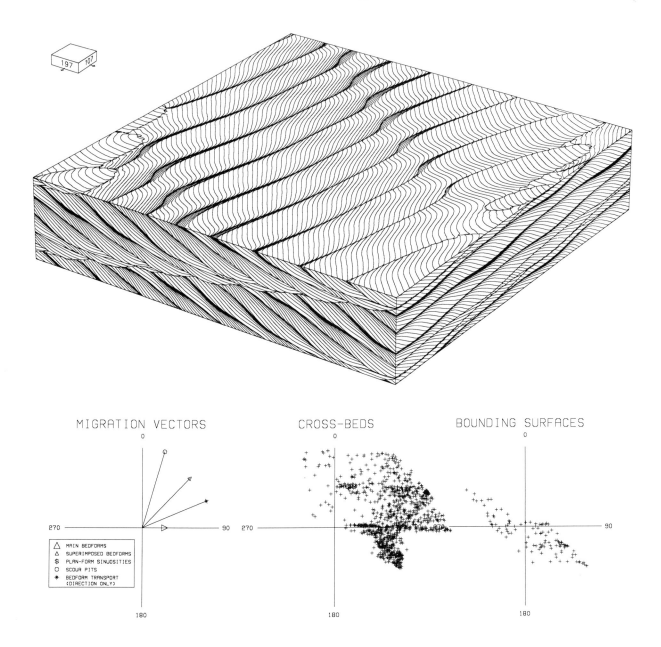

FIG. 72.— Structure formed by straight-crested bedforms with superimposed, sinuous, out-of-phase bedforms migrating obliquely downslope.

RECOGNITION: This structure is similar to that in Figure 46A, but the superimposed bedforms in this example are three-dimensional rather than two-dimensional. One effect of this three-dimensionality is to cause the bed morphology to change through time. The troughs of the main bedforms deepen locally where and when the scour pits in the superimposed bedforms are situated in the main trough. In contrast, an assemblage of two-dimensional bedforms with superimposed two-dimensional bedforms (not parallel to the main bedforms) does not change through time (Fig. 46). Instead, the assemblage of bedforms merely moves through space. Additional effects of the three-dimensionality of the superimposed bedforms are to make individual cross-beds more sinuous and to cause the trough-shaped sets to change in geometry from one location to another.

ORIGIN: Migration of the three-dimensional superimposed bedforms over the main bedforms causes the overall morphology of the bed to change through time, regardless of whether or not the flow changes.

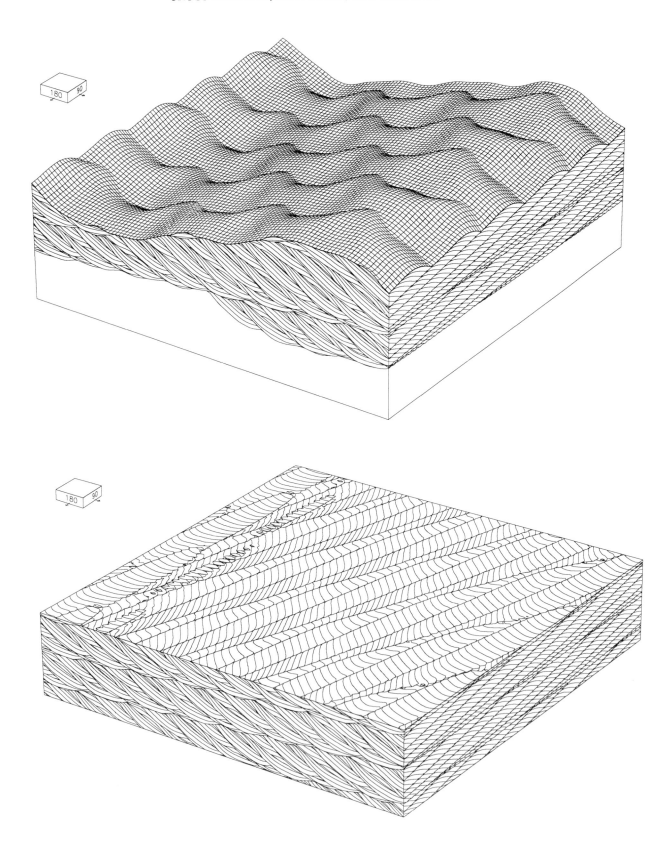

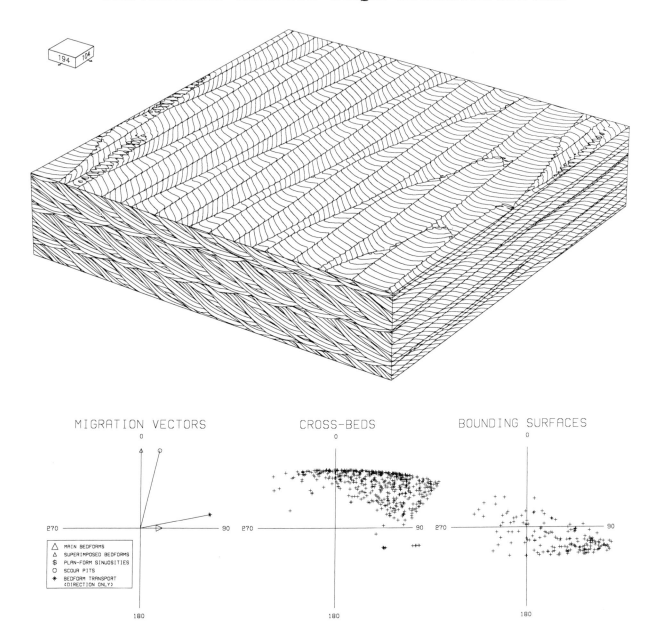

FIG. 73.— Structure formed by bedforms with sinuous, out-of-phase bedforms migrating alongslope. This depositional situation resembles that shown in Figure 72, but here the superimposed bedforms are migrating alongslope rather than obliquely downslope. **RECOGNITION:** This example is so similar to the preceding example (superimposed bedforms migrating obliquely downslope) that the differences probably could not be interpreted in natural exposures. **ORIGIN:** This structure has essentially the same origin as the examples in Figure 46, except that in this example the superimposed bedforms are three-dimensional.

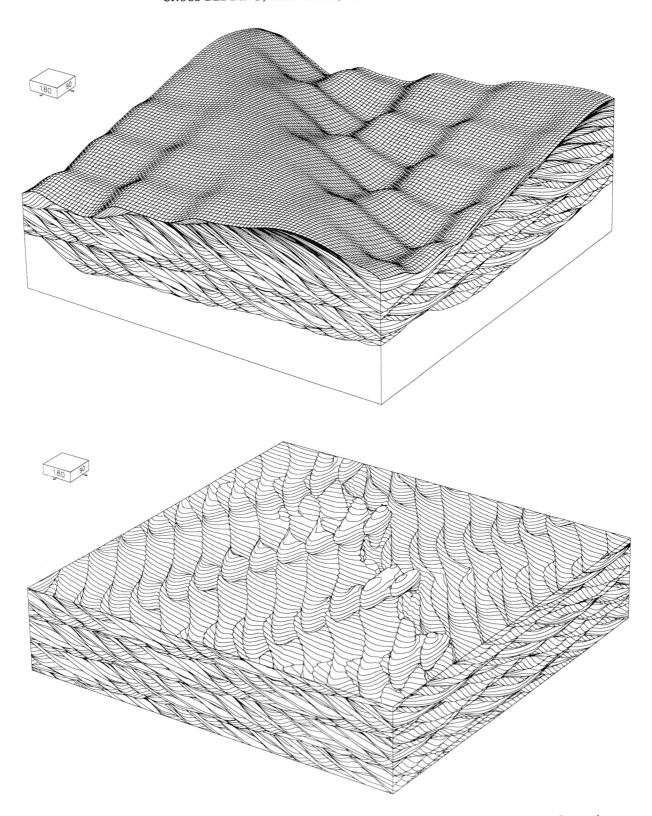

FIG. 74.— Structure deposited by straight-crested bedforms with two sets of straight-crested superimposed bedforms. The superimposed bedforms fluctuate in height, one set growing while the other set diminishes. The periods of height fluctuations are such that the superimposed bedforms experience 13 complete cycles during the time that the main bedform migrates one wavelength. Both sets of

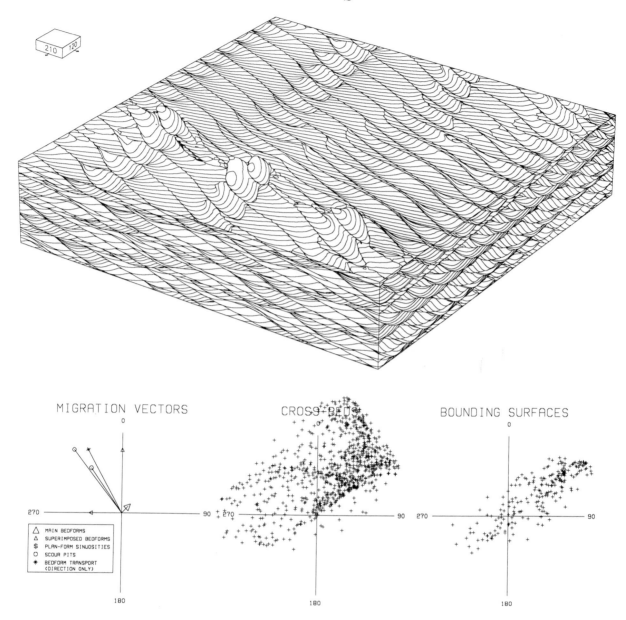

superimposed bedforms are migrating away from the viewer, one set obliquely up the lee slope and one set obliquely downslope. At the time that the bedform is depicted, the bedforms migrating up the main lee slope are decreasing in height, and the bedforms migrating down are increasing in height.

RECOGNITION: This example was created to determine the origin of the real structure shown in Figure 75. The upslope and downslope apparent dip directions of individual cross-beds in the outcrop were a clue that the structure was produced by upslope and downslope migration of superimposed bedforms or by along-crest migration of three-dimensional bedforms. The depositional situation shown here was developed by trial-and-error

experimentation with different depositional situations that satisfied one of these conditions.

ORIGIN: The depositional situation simulated in this computer image requires reversing transport directions and an along-crest (longitudinal) component of sediment transport that is nearly as large as the across-crest (transverse) component of transport. The calculated trend of the transport vector relative to the main crestline is $54°$, which means that the main bedform is slightly more transverse than longitudinal. An example of eolian dunes with this kind of morphology is shown in Figure 76, but bedforms with similar morphology and behavior could also be expected to occur in tidal flows.

FIG. 75.— Structure produced by bedforms like those simulated in Figure 74; Navajo Sandstone (Upper Triassic? and Jurassic) at Johnson Canyon, Utah. The set of cross-beds analagous to the structure in Figure 74 is the compound set that occupies much of the lower half of the photograph; the thickness of that set is approximately 10 m.

RECOGNITION: The main dune migrated from left to right, and, if the computer simulation in Figure 74 is an accurate analogy, the superimposed bedforms migrated obliquely upslope, obliquely downslope, and through the outcrop plane. The three-dimensional structure of this outcrop was not studied in detail in the field; consequently, similarity with the computer simulation in other dimensions (and thus true similarity with the computer simulation) has not been demonstrated.

FIG. 76.— Air photograph of eolian dunes in the Namib desert, Namibia, photographed by Tad Nichols. Like the dunes that are inferred to have produced the bedding in Figure 75, these Namib dunes have two sets of dunes superimposed at approximately 45° to the main dunes. The set that is best developed is migrating from right to left and toward the viewer; a second set is migrating from left to right and toward the viewer. Height of the main dunes is approximately 100 m.

171

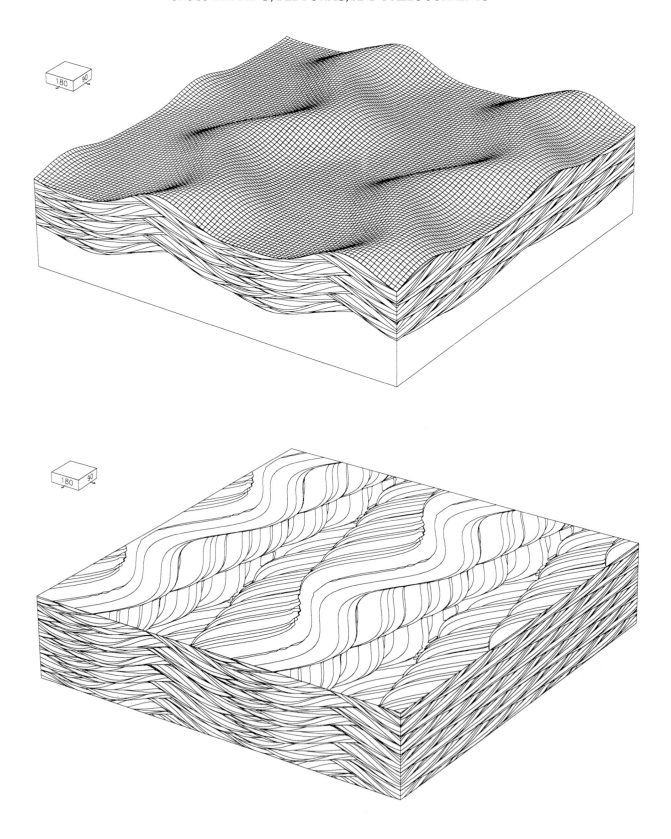

FIG. 77.— Structure produced by reversing, nonmigrating (perfectly longitudinal) bedforms with sinuosities that migrate along-crest. This structure is a reversing analog of the structure illustrated in Figure 55.

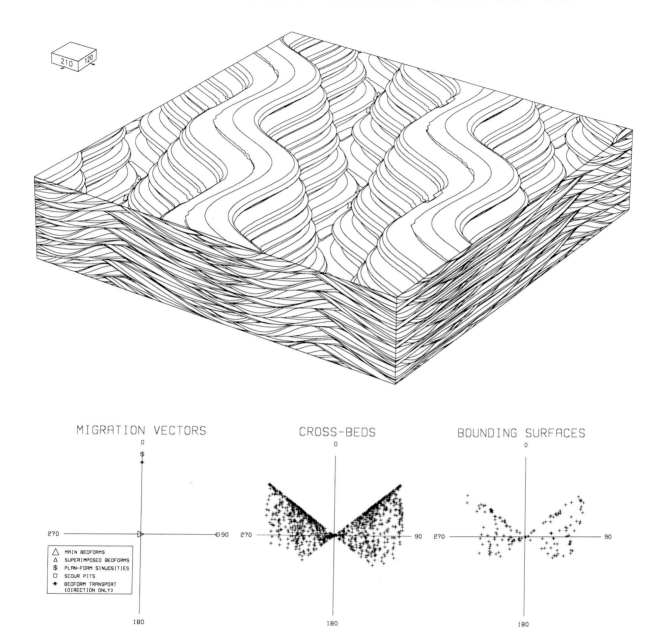

MIGRATION VECTORS

△ MAIN BEDFORMS
△ SUPERIMPOSED BEDFORMS
$ PLAN-FORM SINUOSITIES
O SCOUR PITS
✳ BEDFORM TRANSPORT
  (DIRECTION ONLY)

CROSS-BEDS

BOUNDING SURFACES

RECOGNITION: Vertical climbing of these bedforms is obvious in any section except those parallel to the bedform trend. Along-crest migration of crestline sinuosities induces an along-crest component of dip to the cross-beds (visible in the horizontal section and in sections parallel to the bedform crestline). Bounding surfaces in this structure are produced by two processes: along-crest migration of crestline sinuosities and reversals in bedform asymmetry. Along-crest migration of sinuosities produces trough-shaped bounding surfaces with axes that trend parallel to the bedform crestline. Reversals in bedform asymmetry cause the bounding surfaces to be scallop-shaped in horizontal section; each reversal in asymmetry produces one scallop along the bounding surface. Because of the complexity of these two processes, dips of bounding-surface planes plot in scatter patterns.

ORIGIN: This kind of structure is produced by reversing longitudinal bedforms. Reversing longitudinal dunes with migrating crestline sinuosities have been described by Tsoar (1982, 1983), and their reported internal structure is roughly similar to that shown here. Other possible origins include symmetrical tidal sand waves and ridges, and ripples in oscillatory flows that reverse by less than $180°$.

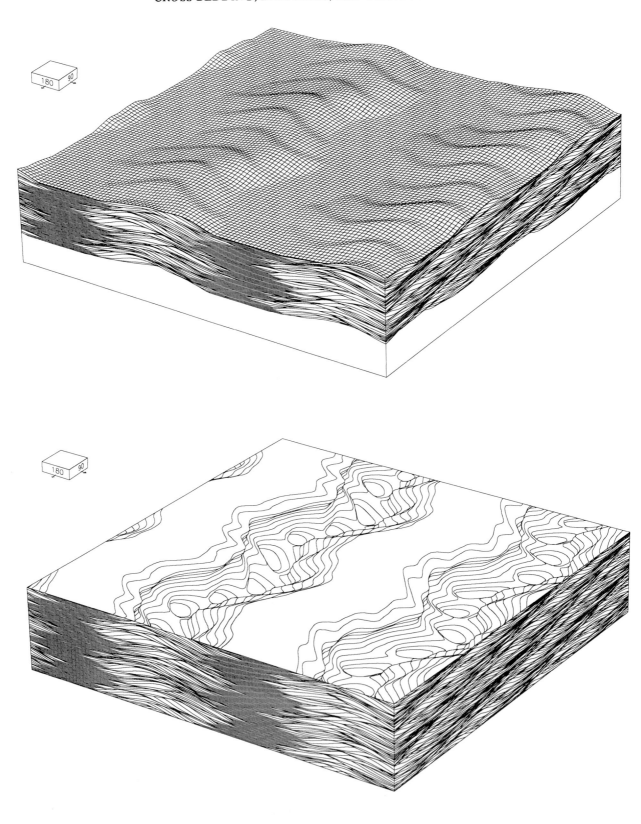

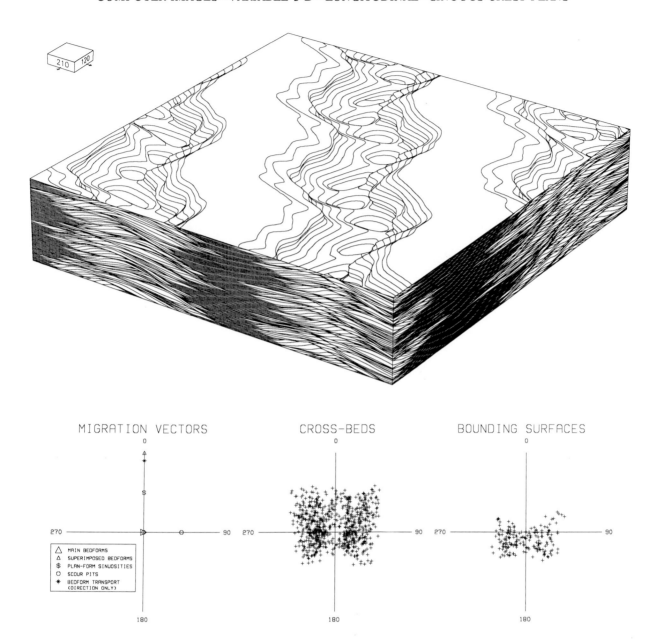

FIG. 78.— Simulation of perfectly longitudinal bedforms with sinuous plan forms, superimposed peaks, and planar troughs. The bedforms reverse in asymmetry; plan-form sinuosities and peaks migrate along-crest.

RECOGNITION: Perfect alignment of the bedforms with the resultant transport direction is demonstrated by vertical accretion of the main bedforms. Reversals in asymmetry of the main bedforms cause the foresets that are deposited on the bedform flanks to intertongue with horizontal beds deposited in the troughs.

ORIGIN: The bedform morphology simulated here is a common morphology of linear dunes. The vertical accretion that is simulated requires unusually perfect alignment of the bedforms with the resultant transport direction.

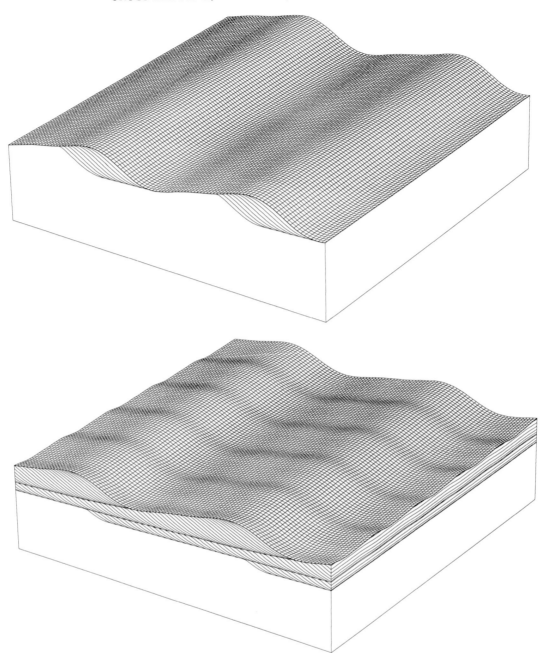

FIG. 79.— A sequence of four images showing the structure produced when one set of bedforms is replaced by smaller bedforms with a different trend. Bedforms in the first set become smaller and migrate more slowly, while those in the second set grow larger and migrate faster.

RECOGNITION: All of the preceding computer-generated structures show depositional situations in which the main bedforms are maintained through time. In real flows, however, one set of bedforms can be replaced by another unrelated set. This example shows the structures that are produced when one set of bedforms is replaced by a second set with a different size and trend. Migration of the second set over the first can partially preserve the first set, thereby enabling recognition of the depositional surface and determination of the spacing of the individual ripples (Fig. 6). The structure that is shown here could probably be recognized in well exposed outcrops, because the sequence begins and ends with bedding that is extremely simple (invariable two-dimensional cross-bedding). Where one complex structure is replaced by another, however, the combination may be so complex as to be virtually impossible to interpret.

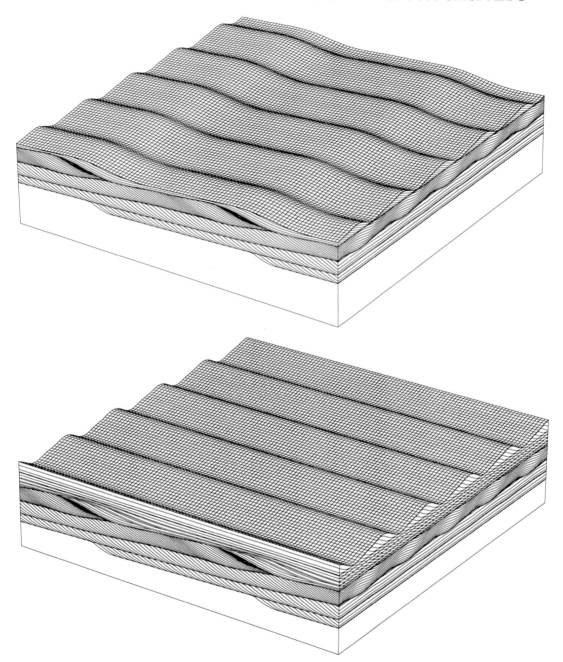

ORIGIN: This structure results from a change in flow direction. The change in flow direction destroys one set of bedforms and creates another. At the start and end of this sequence, the direction of sediment transport cannot be determined; the bedforms that existed at those times were so two-dimensional that they give no indication of along-crest transport. Throughout the majority of the structure, however, the intersecting bedforms can be used to quantify sediment transport in both along-crest and across-crest directions.

Shortly after the start of this sequence, the superimposed bedforms were small and moving slowly (relative to the main bedforms), indicating that the main bedforms were roughly transverse to transport. At the time of the third illustration in this sequence, the second set of bedforms had become larger and were moving faster, and the combination of bedforms and the resulting internal structures resemble examples that were previously shown in steady depositional situations (Fig. 46). A rotation in the transport direction is one mechanism for causing one set of bedforms to become oblique and for eventually causing the creation of a new set of bedforms. Such changes undoubtedly occur in virtually all depositional environments.

CONCLUSIONS

Bedforms and cross-bedding can be simulated mathematically using sine curves. Simulated structures, like real structures, fall into four categories: invariable two-dimensional bedforms and cross-bedding, variable two-dimensional bedforms and cross-bedding, invariable three-dimensional bedforms and cross-bedding, and variable three-dimensional bedforms and cross-bedding. The structures in each category can be distinguished in three-dimensional outcrops and in polar plots of cross-beds and bounding surfaces, but the structures often are indistinguishable in two-dimensional exposures.

Variability of bedforms can result from two processes: flow fluctuations that cause systematic changes to entire populations of bedforms, and migration of superimposed or intersecting bedforms in flows that may be steady. Cross-bedding produced by these two processes can often be distinguished by the relations between the dip directions of cross-beds and bounding surfaces within the cross-stratified beds. Structures produced by bedforms that change morphology or path of climb in response to flow fluctuations have cross-beds and bounding surfaces with the same strike. Structures produced by superimposed bedforms have cross-beds and bounding surfaces with differing strikes, because the crestlines of the superimposed bedforms are unlikely to parallel the crestlines of the main bedforms exactly.

Cyclic cross-bedding produced by cyclic flow fluctuations is useful for determining flow velocities. Cyclic cross-bedding produced by superimposed bedforms is useful for determining flow directions; structures deposited by bedforms that were oriented obliquely to the sediment transport direction can be recognized by along-crest migration of superimposed bedforms.

Cross-bedding provides an extensive and readily observable record of bedform behavior. Most studies of cross-bedding have been directed at interpreting ancient deposits, but study of cross-bedding also provides useful information about how modern bedforms behave. For example, evidence of along-crest migration of superimposed bedforms is common in the geologic record and indicates that oblique bedforms are more common than is generally appreciated. Similarly, the importance of coexisting bedforms with differing orientations has not been appreciated from studies of modern bedforms, despite the abundance of such bedforms. This use of cross-bedding—to study the behavior of modern bedforms—has been a neglected field of sedimentology.

Computer-graphics modeling of cross-bedding is useful not only for illustrative purposes, as in this publication, but also as a research tool. By trial-and-error simulation of specific structures, the morphology and behavior of the bedforms that produced the bedding can be quantitatively reconstructed.

APPENDIX A

INPUT PARAMETERS USED TO CREATE COMPUTER-GENERATED CROSS-BEDDING
(VALUES ARE THOSE USED IN FIGURE 73; UNITS ARE DEFINED BELOW)

MAIN BEDFORMS

100.0	Spacing
0.0	Phase (controls placement of the bedforms within the block diagram)
1.0	Symmetry (0=symmetric; +1=asymmetric; -1=reversed)
0.0	Magnitude of symmetry fluctuations (same units as symmetry)
1.0	Period of symmetry fluctuations
0.0	Initial phase in symmetry cycle
1.0	Mean steepness (a value of 1 gives a height/steepness ratio of 1/15)
0.0	Magnitude of steepness fluctuations
1.0	Period of steepness fluctuations
0.0	Initial phase of steepness fluctuations
0.0	Spacing (along-crest) of first set of sinuosities
0.0	Amplitude (in a horizontal plane) of sinuosities in first set
0.0	Phase of sinuosities in first set (controls position along-crest)
0.0	Migration speed (along-crest) of sinuosities in first set
0.0	Spacing (along-crest) of second set of sinuosities
0.0	Amplitude (in a horizontal plane) of sinuosities in second set
0.0	Phase of sinuosities in second set (controls position along-crest)
0.0	Migration speed (along-crest) of sinuosities in second set
90.0	Migration direction
0.5	Migration speed
0.0	Magnitude of speed fluctuations
1.0	Period of speed fluctuations
0.0	Initial phase in migration-speed cycle

FIRST SET OF SUPERIMPOSED BEDFORMS

37.5	Spacing
180	Phase (controls placement of the bedforms within the block diagram)
1.0	Symmetry (0=symmetric; +1=asymmetric; -1=reversed)
0.0	Magnitude of symmetry fluctuations
1.0	Period of symmetry fluctuations
0.0	Initial phase in symmetry cycle
0.8	Mean steepness (a value of 1 gives a height/steepness ratio of 1/15)
0.0	Magnitude of steepness fluctuations

1.0	Period of steepness fluctuations
0.0	Initial phase of steepness fluctuations
25.0	Spacing (along-crest) of first set of sinuosities
3.3	Amplitude (in a horizontal plane) of sinuosities in first set
180.0	Phase of sinuosities in first set (controls position along-crest)
0.0	Migration speed (along-crest) of sinuosities in first set
12.5	Spacing (along-crest) of second set of sinuosities
1.0	Amplitude (in a horizontal plane) of sinuosities in second set
270.0	Phase of sinuosities in second set (controls position along-crest)
0.0	Migration speed (along-crest) of sinuosities in second set
0.0	Migration direction
2.0	Migration speed
0.0	Magnitude of speed fluctuations
1.0	Period of speed fluctuations
0.0	Initial phase in migration-speed cycle

SECOND SET OF SUPERIMPOSED BEDFORMS

37.5	Spacing
0.0	Phase (controls placement of the bedforms within the block diagram)
1.0	Symmetry (0=symmetric; +1=asymmetric; -1=reversed)
0.0	Magnitude of symmetry fluctuations
1.0	Period of symmetry fluctuations
0.0	Initial phase in symmetry cycle
0.8	Mean steepness (a value of 1 gives a height/steepness ratio of 1/15)
0.0	Magnitude of steepness fluctuations
1.0	Period of steepness fluctuations
0.0	Initial phase of steepness fluctuations
25.0	Spacing (along-crest) of first set of sinuosities
3.3	Amplitude (in a horizontal plane) of sinuosities in first set
0.0	Phase of sinuosities in first set (controls position along-crest)
0.0	Migration speed (along-crest) of sinuosities in first set
12.5	Spacing (along-crest) of second set of sinuosities

1.0 Amplitude (in a horizontal plane) of sinuosities in second set

270.0 Phase of sinuosities in second set (controls position along-crest)

0.0 Migration speed (along-crest) of sinuosities in second set

0.0 Migration direction

2.0 Migration speed

0.0 Magnitude of speed fluctuations

1.0 Period of speed fluctuations

0.0 Initial phase in migration-speed cycle

BEDFORM SUPERPOSITIONING

5 Type of superpositioning (integer value from 1 to 6, see below)

-1.0 Elevation of interdune flats (see below)

DEPOSITION

0.04 Rate of deposition

0.0 Magnitude of fluctuations in rate of deposition

1.0 Period of fluctuations in rate of deposition

0.0 Initial phase in cycle of deposition

TIME INTERVALS REPRESENTED

300 Time of the beginning of depositional episode

1 Time of the end of depositional episode

1 Interval between drawing of cross-beds

PLOTTING AND ANNOTATION PARAMETERS

1 Number of frames in sequence (normally is 1, but can be more for animated sequences)

- Caption

- Input file

- Frame

EXPLANATIONS AND UNITS OF PARAMETERS

Length: all length dimensions are defined relative to the lengths of the sides of the block, which are 100 units long.

Phase: all phases are given in degrees and describe the situation at t=zero.

Symmetry: dimensionless.

Time: arbitrary units that describe all parameters of time (periods of cyclicity, migration speeds, and deposition rate).

Direction: in degrees, oriented as indicated in the computer images.

Bedforms are superimposed following one of six rules:

(1) superimposed bedforms are placed on the main bedforms by simple addition (Fig. 46A);
(2) height of the superimposed bedforms is proportional to the local elevation of the main bedform (Fig. 38);
(3) heights of all bedforms are calculated separately, and the elevation at any point on the surface is chosen to be that of whatever bedform is locally highest (Fig. 34);
(4) height of the superimposed bedforms is inversely proportional to the local elevation of the main bedform (Fig. 78);
(5) two sets of superimposed bedforms are created as in (3) and then added to the main bedforms as in (1) (Fig. 65);
(6) the first set of superimposed bedforms is incorporated as in (2), and the second set is added as in (1).

Elevation of interdune flats: a value of less than -1.0 will generally produce no interdune flats; the flats in Figure 78 were defined with a value of -0.15.

In any one depositional situation, many of the parameters listed have no effect on the resulting structure. In this example, bedform steepness is constant through time, and the parameters that define the period and initial phase of steepness fluctuations have no effect on the results.

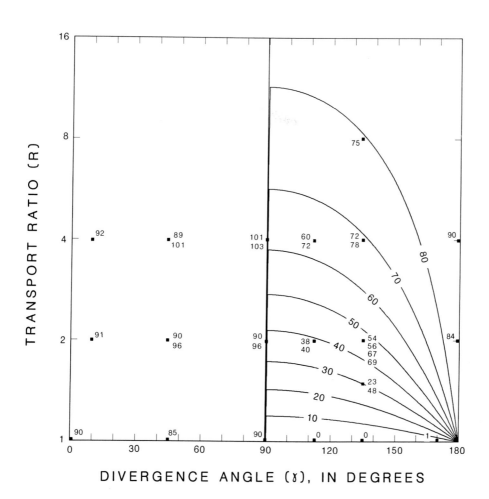

FIG. B-1.— Plot of bedform alignment relative to the resultant transport direction in bidirectional flows. The plot is based on an experimental study of wind ripples (Rubin and Hunter, 1987) and shows bedform alignment as a function of two variables: the divergence angle (the angle between the two transport directions) and the transport ratio (the ratio of the amount of transport in the dominant direction to the amount in the subordinate direction). Values plotted at the points are the alignments of the experimental bedforms. Transverse bedforms (alignments of roughly 90°) were generally produced when the divergence angle was less than 90°, when the transport ratio was large, or when the divergence angle approached 180°; longitudinal bedforms (alignments of roughly 0°) were produced when the divergence angle was between 90° and 180°, and the transport ratio approached unity; oblique bedforms (intermediate alignments) were produced when the divergence angle was between 90° and 180°, and the transport ratio was between unity and approximately eight. The contours are the bedform alignments that were calculated to yield the maximum gross bedform-normal transport (transport in which transports across the bedform crestline are summed as two positive numbers). The agreement of the observed and calculated alignments suggests that the experimental bedforms follow this rule of alignment. The fact that the experimental bedforms all follow the same rule of alignment suggests that they do not require differing flow dynamics for their origin (that is, they are basically the same kind of bedform).

APPENDIX C

FIGURE	DIMENSIONALITY OF ASSEMBLAGE	VARIABILITY OF ASSEMBLAGE	PLAN-FORM SINUOSITIES	SUPERIMPOSED BEDFORMS	ORIENTATION TO TRANSPORT	UNSTEADY FLOW REQUIRED?	BEDDING FEATURES
3	2	INVARIABLE	NONE	NONE	UNDEFINABLE	NO	RIPPLEFORM
4	2	INVARIABLE	NONE	NONE	UNDEFINABLE	NO	RIPPLEFORM
5	2	INVARIABLE	NONE	NONE	UNDEFINABLE	NO	TABULAR SETS
8	2	INVARIABLE	NONE	NONE	UNDEFINABLE	NO	NONDEPOSITIONAL
9	2	INVARIABLE	NONE	NONE	UNDEFINABLE	NO	NONDEPOSITIONAL
10	2	INVARIABLE	NONE	NONE	UNDEFINABLE	NO	NONDEPOSITIONAL
13	2	VARIABLE	NONE	NONE	UNDEFINABLE	YES	SIGMOIDAL SETS
15	2	VARIABLE	NONE	NONE	UNDEFINABLE	YES	UNDULATING SET BOUNDARIES
16	2	VARIABLE	NONE	NONE	UNDEFINABLE	YES	SCALLOPS
17	2	VARIABLE	NONE	NONE	UNDEFINABLE	YES	SCALLOPS
18	2	VARIABLE	NONE	NONE	UNDEFINABLE	YES	ZIG-ZAGS
19	2	VARIABLE	NONE	NONE	UNDEFINABLE	YES	ZIG-ZAGS
21	2	VARIABLE	NONE	NONE	UNDEFINABLE	YES	SCALLOPS
22	2	VARIABLE	NONE	NONE	UNDEFINABLE	YES	SCALLOPS
25	2	VARIABLE	NONE	STRAIGHT	UNDEFINABLE	NO	CYCLIC FORESETS
27	2	VARIABLE	NONE	STRAIGHT	UNDEFINABLE	NO	CYCLIC FORESETS
29	2	VARIABLE	NONE	STRAIGHT	UNDEFINABLE	YES	SCALLOPS
32	3	INVARIABLE	IN-PHASE	NONE	TRANSVERSE	NO	NEARLY PLANAR SET BOUNDARIES
34	3	INVARIABLE	OUT-OF-PHASE	NONE	TRANSVERSE	NO	TROUGH SETS
36	3	INVARIABLE	PSEUDORANDOM	NONE	TRANSVERSE	NO	TROUGH SETS
38	3	INVARIABLE	NONE	NONE	TRANSVERSE	NO	UNDULATING SET BOUNDARIES
40	3	INVARIABLE	NONE	NONE	TRANSVERSE	NO	RIPPLEFORM
42	3	INVARIABLE	IN-PHASE	NONE	OBLIQUE	NO	TROUGH SETS
43	3	INVARIABLE	OUT-OF-PHASE	NONE	OBLIQUE	NO	TROUGH SETS
45	3	INVARIABLE	PSEUDORANDOM	NONE	OBLIQUE	NO	TROUGH SETS
46	3	INVARIABLE	NONE	STRAIGHT	OBLIQUE	NO	SCALLOPS
55	3	INVARIABLE	IN-PHASE	NONE	LONGITUDINAL	NO	ZIG-ZAGS
56	3	INVARIABLE	NONE	SINUOUS	LONGITUDINAL	NO	TROUGH SUBSETS
58	3	VARIABLE	IN-PHASE	NONE	TRANSVERSE	YES	UNDULATING AND SCALLOPED
59	3	VARIABLE	NONE	STRAIGHT	TRANSVERSE	YES	ZIG-ZAGS, SCALLOPS, AND TROUGH SETS
63	3	VARIABLE	NONE	STRAIGHT	TRANSVERSE	YES	RIPPLEFORM
65	3	VARIABLE	IN-PHASE	SINUOUS	TRANSVERSE	NO	TROUGH SUBSETS
66	3	VARIABLE	IN-PHASE	STRAIGHT	TRANSVERSE	NO	UNDULATING SET BOUNDARIES
67	3	VARIABLE	IN-PHASE	STRAIGHT	TRANSVERSE	YES	DOUBLE CYCLICITY
69	3	VARIABLE	IN-PHASE	NONE	OBLIQUE	YES	SCALLOPS
71	3	VARIABLE	NONE	STRAIGHT	OBLIQUE	YES	IRREGULAR SCALLOPS
72	3	VARIABLE	NONE	SINUOUS	OBLIQUE	NO	IRREGULAR SCALLOPS
73	3	VARIABLE	NONE	SINUOUS	OBLIQUE	NO	IRREGULAR SCALLOPS
74	3	VARIABLE	NONE	2 SETS	OBLIQUE	YES	IRREGULAR SCALLOPS
77	3	VARIABLE	IN-PHASE	NONE	LONGITUDINAL	YES	ZIG-ZAGS
78	3	VARIABLE	IN-PHASE	PEAKS AND SADDLES	LONGITUDINAL	YES	IRREGULAR ZIG-ZAGS
79	3	VARIABLE	IN-PHASE	YES	LONGITUDINAL	YES	STRUCTURE CHANGES THROUGH TIME

CROSS-REFERENCE TABLE OF COMPUTER-GENERATED STRUCTURES

REFERENCES

ALLEN, J.R.L., 1968, Current Ripples; Their Relation to Patterns of Water and Sediment Motion: North-Holland Publishing Company, Amsterdam, 433 p.

———, 1969, On the geometry of current ripples in relation to stability of fluid flow: Geografiska Annaler, v. 51A, p. 61-96.

———, 1972, A theoretical and experimental study of climbing ripple cross lamination, with a field application to the Uppsala Esker: Geografiska Annaler, v. 53A, p. 157-187.

———, 1973, Features of cross-stratified units due to random and other changes in bedforms: Sedimentology, v. 20, p. 189-202.

———, 1977, The plan shape of current ripples in relation to flow conditions: Sedimentology, v. 24, p. 53-62.

———, 1978, Polymodal dune assemblages: an interpretation in terms of dune creation-destruction in periodic flows: Sedimentary Geology, v. 20, p. 17-28.

———, 1979, A model for the interpretation of wave ripple-marks using their wavelength, textural composition, and shape: Geological Society of London Proceedings, v. 136, p. 673-682.

———, 1981, Paleotidal speeds and ranges estimated from cross-bedding sets with mud drapes: Nature, v. 293, p. 394-396.

———, 1982, Sedimentary Structures: Their Character and Physical Basis: Elsevier, New York, v. I, 594 p., v. II, 644 p.

ALLEN, P.A., AND HOMEWOOD, P., 1984, Evolution and mechanics of a Miocene tidal sand wave: Sedimentology, v. 31, p. 63-81.

ASHLEY, G.M., SOUTHARD, J.B., AND BOOTHROYD, J.C., 1982, Deposition of climbing-ripple beds: a flume simulation: Sedimentology, v. 29, p. 67-79.

BAGNOLD, R.A., 1941, The Physics of Blown Sand and Desert Dunes: Methuen, London, 265 p.

———, 1946, Motion of waves in shallow water, interactions between waves and sand bottoms: Royal Society of London Proceedings, Series A, v. 187, p. 1-18.

BANKS, N.L., 1973, The origin and significance of some downcurrent-dipping cross-stratified sets: Journal of Sedimentary Petrology, v. 43, p. 423-427.

———, AND COLLINSON, J.D., 1975, The size and shape of small-scale current ripples: an experimental study: Sedimentology, v. 22, p. 583-599.

BENEDICT, P.C., ALBERTSON, M.L., AND MATEJKA, D.Q., 1955, Total sediment load measured in a turbulence flume: American Society of Civil Engineers Transactions, v. 120, p. 457-489.

BEUTNER, F.C., FLUECKINGER, L.A., AND GARD, T.M., 1967, Bedding geometry in a Pennsylvanian channel sandstone: Geological Society of America Bulletin, v. 78, p. 911-916.

BOERSMA, J.R., 1969, Internal structure of some tidal megaripples on a shoal in the Westerschelde estuary, The Netherlands: Geologie en Mijnbouw, v. 48, p. 409-414.

———, AND TERWINDT, J.H.J., 1981, Neap-spring tide sequences of intertidal shoal deposits in a mesotidal estuary: Sedimentology, v. 28, p. 151-170.

———, VAN DE MEENE, E.A., AND TJALSMA, R.C., 1968, Intricated cross-stratification due to interaction of a mega ripple with its leeside of backflow ripples (upper-pointbar deposits, Lower Rhine)[sic]: Sedimentology, v. 11, p. 147-162.

BOHACS, K.M., 1981, Flume experiments on the kinematics and dynamics of large-scale bed forms: Unpublished Ph.D. Dissertation, Massachusetts Institute of Technology, Cambridge, Massachusetts, 177 p.

BREED, C.S., AND BREED, J.W., 1979, Dunes and other windforms in central Australia (and a comparison with linear dunes on the Moenkopi Plateau, Arizona), in El-Baz, F., and Warner, D.M., eds., Apollo-Soyuz Test Project Summary Science Report: National Aeronautics and Space Administration SP-412, v. 2, p. 319-358.

BROOKFIELD, M.E., 1977, The origin of bounding surfaces in ancient eolian sandstones: Sedimentology, v. 24, p. 303-332.

CLIFTON, H.E., 1976, Wave-formed sedimentary structures -- a conceptual model: Society of Economic Paleontologists and Mineralogists Special Publication 24, p. 126-148.

COLEMAN, J.M., 1969, Brahmaputra River: channel processes and sedimentation: Sedimentary Geology, v. 13, p. 129-239.

COSTELLO, W.R., AND SOUTHARD, J.B., 1981, Flume experiments on lower-flow-regime bed forms in coarse sand: Journal of Sedimentary Petrology, v. 51, p. 849-864.

DALRYMPLE, R.W., 1984, Morphology and internal structure of sandwaves in the Bay of Fundy: Sedimentology, v. 31, p. 365-382.

———, KNIGHT, R.J., AND LAMBIASE, J.J., 1978, Bedforms and their hydraulic stability relationships in a tidal environment, Bay of Fundy, Canada: Nature, v. 275, p. 100-104.

DAVIES, T.R.H., 1982, Bed shear stress over subaqueous dunes, and the transition to upper-stage plane beds—discussion: Sedimentology, v. 29, p. 743-744.

DIETRICH, W.R, AND SMITH, J.D., 1984, Bedload transport in a river meander: Water Resources Research, v. 20, p. 1355-1380.

DINGLER, J.R., 1974, Wave-formed ripples in nearshore sands: Unpublished Ph.D. Dissertation, University of California, San Diego, California, 136 p.

ELLIOTT, T., AND GARDINER, A.R., 1981, Ripple, megaripple and sandwave bedforms in the macro-tidal Loughor Estuary, South Wales, U.K., *in* Nio, S.-D., Shuttenhelm, R.T.E., and van Weering, Tj.C.E., eds., Holocene Marine Sedimentation in the North Sea Basin: International Association of Sedimentologists Special Publication 5, Blackwell Scientific Publications, Oxford, p. 51-64.

FOLK, R.L., 1971, Longitudinal dunes of the northwestern edge of the Simpson Desert, Northern Territory, Australia: 1. Geomorphology and grain-size relationships: Sedimentology, v. 16, p. 5-54.

FRYBERGER, S.G., 1979, Dune forms and wind regime, *in* McKee, E.D., ed., A Study of Global Sand Seas: U.S. Geological Survey Professional Paper 1052, p. 137-170.

GILBERT, G.K., 1914, The transportation of debris by running water: U.S. Geological Survey Professional Paper 86, 263 p.

GONCHAROV, V.N., 1929, The movement of sediment on the channel bottom: Sci. Inst. Amelioration, N. Caucasus Br., Translation No. 44, Mechanical Engineering Department, University of California, Berkeley, California, p. 851-920.

GUSTAVSON, T.C., ASHLEY, G.M., AND BOOTHROYD, J.C., 1975, Depositional sequences in glaciolacustrine deltas, *in* MacDonald, B.C., and Jopling, A.V., eds., Glaciofluvial and Glaciolacustrine Sedimentation: Society of Economic Paleontologists and Mineralogists Special Publication 23, p. 264-280.

GUY, H.P., SIMONS, D.B., AND RICHARDSON, E.V., 1966, Summary of alluvial channel data from flume experiments, 1956-1961: U.S. Geological Survey Professional Paper 462-I, 96 p.

HAND, B.M., 1974, Supercritical flow in density currents: Journal of Sedimentary Petrology, v. 44, p. 637-648.

HARMS, J.C., 1969, Hydraulic significance of some sand ripples: Geological Society of America Bulletin, v. 80, p. 363-396.

———, SOUTHARD, J.B., AND WALKER, 1982, Structures and Sequences in Clastic Rocks: Society of Economic Paleontologists and Mineralogists Short Course No. 9, Tulsa, Oklahoma, 394 p.

HEMINGWAY, J.E., AND CLARKE, A.M., 1963, Structure of ripple marks: Nature, v. 189, p. 847-850.

HEREFORD, R., 1977, Deposition of the Tapeats Sandstone (Cambrian) in central Arizona: Geological Society of America Bulletin, v. 88, p. 199-211.

HUBBELL, D.W., 1964, Apparatus and techniques for measuring bedload: U.S. Geological Survey Water-Supply Paper 1748, 74 p.

HUNTER, R.E., 1977, Terminology of cross-stratified sedimentary layers and climbing ripple structures: Journal of Sedimentary Petrology, v. 47, p. 697-706.

———, 1985, Subaqueous sand-flow cross strata: Journal of Sedimentary Petrology, v. 55, p. 886-894.

———, RICHMOND, B.M., AND ALPHA, T.R., 1983, Storm-controlled oblique dunes of the Oregon coast: Geological Society of America Bulletin, v. 94, p. 1450-1465.

———, AND RUBIN, D.M., 1983, Interpreting cyclic crossbedding, with an example from the Navajo Sandstone, *in* Brookfield, M.E., and Ahlbrandt, T.S., eds., Eolian Sediments and Processes: Elsevier, Amsterdam, p. 429-454.

INMAN, D.L., 1957, Wave-generated ripples in nearshore sands: U.S. Army Corps of Engineers, Beach Erosion Board Technical Memorandum No. 100, 66 p.

JOPLING, A.V., 1965, Laboratory study of the distribution of grain sizes in cross-bedded deposits, *in* Middleton, G.V., ed., Primary Sedimentary Structures and their Hydrodynamic Interpretation: Society of Economic Paleontologists and Mineralogists Special Publication 12, p. 53-65.

———, AND WALKER, R.G., 1968, Morphology and origin of ripple-drift cross-lamination, with examples from the Pleistocene of Massachusetts: Journal of Sedimentary Petrology, v. 38, p. 971-984.

KENNEDY, J.F., 1969, The formation of sediment ripples, dunes, and antidunes: Annual Review of Fluid Mechanics, v. 16, p. 147-168.

KOMAR, P.D., 1973, An occurrence of "brick pattern" oscillatory ripple marks at Mono Lake, California: Journal of Sedimentary Petrology, v. 43, p. 1111-1113.

———, 1974, Oscillatory ripple marks and the evaluation of ancient wave conditions and environments: Journal of Sedimentary Petrology, v. 44, p. 169-180.

MACHIDA, T., INOKUCHI, M., MATSUMOTO, E., TAKAYUKI, I., AND IKEDA, H., 1974, Sand ripple patterns and their arrangement on the sea bottom of the Tatado beach, Izu Peninsula, central Japan: Science Reports of the Tokyo Kyoiku Daigaku, Section C, v. 12, p. 1-16.

McCABE, P.J., AND JONES, C.M., 1977, Formation of reactivation surfaces within superimposed deltas and bedforms: Journal of Sedimentary Petrology, v. 47, p. 707-715.

McKEE, E.D., 1939, Some types of bedding in the Colorado River delta: Journal of Geology, v. 47, p. 64-81.

———, 1965, Experiments in ripple lamination, *in* Middleton, G.V., ed., Primary Sedimentary Structures and their Hydrodynamic Interpretation: Society of Economic Paleontologists and Mineralogists Special Publication 12, p. 66-83.

MIDDLETON, G.V., AND SOUTHARD J.B., 1984, Mechanics of Sediment Movement: Society of Economic Paleontologists and Mineralogists Short Course No. 3, Providence, Rhode Island, 401 p.

NEWTON, R.S., 1968, Internal structure of wave-formed ripple marks in the nearshore zone: Sedimentology, v. 11, p. 275-292.

RUBIN, D.M., 1987, Formation of scalloped crossbedding without unsteady flows: Journal of Sedimentary Petrology, v. 57, p. 39-45.

———, AND HUNTER, R.E., 1982, Bedform climbing in theory and nature: Sedimentology, v. 29, p. 121-138.

———, AND ———, 1983, Reconstructing bedform assemblages from compound crossbedding, *in* Brookfield, M.E., and Ahlbrandt, T.S., eds., Eolian Sediments and Processes: Elsevier, Amsterdam, p. 407-427.

———, AND ———, 1984, Origin of extensive bedding planes in aeolian sandstones—reply: Sedimentology, v. 31, p. 128-132.

———, AND ———, 1985, Why longitudinal dunes are rarely recognized in the geologic record: Sedimentology, v. 32, p. 147-157.

———, AND ———, 1987, Bedform alignment in directionally varying flows: Science.

———, AND McCULLOCH, D.S., 1980, Single and superimposed bedforms: a synthesis of San Francisco Bay and flume observations: Sedimentary Geology, v. 26, p. 207-231.

SCHMIDT, J.C., 1986, Controls on flow separation and sedimentation in a bedrock river, Colorado River, Grand Canyon: Geological Society of America, Abstracts with Programs, v. 18, p. 741.

SHARP, R.P., 1963, Wind ripples: Journal of Geology, v. 71, p. 617-636.

SIMONS, D.B., RICHARDSON, E.V., AND NORDIN, C.F., Jr., 1965, Bedload equations for ripples and dunes: U.S. Geological Survey Professional Paper 462-H, 9 p.

SMITH, J.D., AND McLEAN, S.R., 1977, Spatially averaged flow over a wavy surface: Journal of Geophysical Research, v. 82, p. 1735-1746.

SORBY, H.C., 1859, On the structures produced by the currents present during the deposition of stratified rocks: The Geologist, v. 2, p. 137-147.

SOUTHARD, J.B., 1971, Representation of bed configurations in depth-velocity-size diagrams: Journal of Sedimentary Petrology, v. 41, p. 903-915.

———, 1975, Bed configurations, *in* Harms, J.C., Southard, J.B., Spearing, D.R., and Walker, R.G., eds., Depositional Environments as Interpreted from Primary Sedimentary Structures and Stratification Sequences: Society of Economic Paleontologists and Mineralogists Short Course No. 2, Tulsa, Oklahoma, p. 5-43.

STEIN, R.A., 1965, Laboratory studies of total and apparent load: Journal of Geophysical Research, v. 70, p. 1831-1842.

STOKES, W.L., 1964, Eolian varving in the Colorado Plateau: Journal of Sedimentary Petrology, v. 34, p. 429-432.

———, 1968, Multiple parallel-truncation bedding planes: a feature of wind-deposited sandstone formations: Journal of Sedimentary Petrology, v. 38, p. 510-515.

STRIDE, A.H., 1965, Preservation of some marine current-bedding: Nature, v. 206, p. 498-499.

TERWINDT, J.H.J., 1981, Origin and sequences of sedimentary structures in inshore mesotidal deposits of the North Sea, *in* Nio, S.-D., Shuttenhelm, R.T.E., and van Weering, Tj.C.E., eds., Holocene Marine Sedimentation in the North Sea Basin: International Association of Sedimentologists Special Publication 5, Blackwell Scientific Publications, Oxford, p. 4-26.

———, AND BROUWER, M.J.N., 1986, The behaviour of intertidal sandwaves during neap-spring tide cycles and the relevance for palaeoflow reconstructions: Sedimentology, v. 33, p. 1-31.

TSOAR, H., 1982, Internal structure and surface geometry of longitudinal (seif) dunes: Journal of Sedimentary Petrology, v. 52, p. 823-831.

———, 1983, Dynamic processes acting on a longitudinal (seif) sand dune: Sedimentology, v. 30, p. 567.

TWIDALE, C.R., 1981, Age and origin of longitudinal dunes in the Simpson and other sand ridge deserts: Die Erde, v. 112, p. 231-247.

VISSER, M.J., 1980, Neap-spring cycles reflected in Holocene sub-tidal large-scale bedform deposits: a preliminary note: Geology, v. 8, p. 543-546.

WALKER, D.W., 1981, An experimental study of wind ripples: Unpublished M.S. Thesis, Massachusetts Institute of Technology, Cambridge, Massachusetts, 145 p.

WALKER, R.G., 1963, Distinctive types of ripple-drift cross-lamination: Sedimentology, v. 2, p. 173-188.

INDEX

(For computer-generated depositional situations, see also Figure 2 and Appendix C.)